GROW YOUR OWN MUSHROOMS

A Beginner's Guide

Quarto.com

First Published in 2024 by New Shoe Press, an imprint of The Quarto Group,
100 Cummings Center, Suite 265-D, Beverly, MA 01915, USA.
T (978) 282-9590 F (978) 283-2742

Essential, In-Demand Topics, Four-Color Design, Affordable Price
New Shoe Press publishes affordable, beautifully designed books covering evergreen, in-demand subjects. With a goal to inform and inspire readers' everyday hobbies, from cooking and gardening to wellness and health to art and crafts, New Shoe titles offer the ultimate library of purposeful, how-to guidance aimed at meeting the unique needs of each reader. Reimagined and redesigned from Quarto's best-selling backlist, New Shoe books provide practical knowledge and opportunities for all DIY enthusiasts to enrich and enjoy their lives.

Visit Quarto.com/New-Shoe-Press for a complete listing of the New Shoe Press books.

New Shoe Press titles are also available at discount for retail, wholesale, promotional, and bulk purchase. For details, contact the Special Sales Manager by email at specialsales@quarto.com or by mail at The Quarto Group, Attn: Special Sales Manager, 100 Cummings Center, Suite 265-D, Beverly, MA 01915, USA.

ISBN: 978-0-7603-9078-8
eISBN: 978-0-7603-9079-5

The content in this book was previously published in *Mushroom Cultivation* (Quarry Books 2018) by Tavis Lynch.

Library of Congress Cataloging-in-Publication Data available

Photography: All photos by Dewitz Photography/Travis Dewitz, except:
Shutterstock: 9, 59, 69, 81, 82 (bottom), 91
Author: 44 (top), 56, 61, 63, 65, 67, 73, 74, 75 (top), 79, 88
Receipe development: Sonia Turek

The information in this book is for educational purposes only. It is not intended to replace the advice of a physician or medical practitioner. Please see your health-care provider before beginning any new health program.

GROW YOUR OWN MUSHROOMS

A Beginner's Guide

An Illustrated Guide to Cultivating Your Own Mushrooms at Home

TAVIS LYNCH

MUSHROOMS

Contents

Introduction

Over the past few years, it seems as if mushrooms are popping up everywhere: on restaurant menus, in grocery aisles, and at local farmers' markets. And not just the ubiquitous white buttons we've seen for years; what once were exotic are now almost commonplace. Shiitake, chanterelle, cremini, enoki—the list grows longer every year. Maybe it's because they're so healthy: mushrooms are just packed with vitamins and antioxidants. Maybe it's a desire for new flavors and textures—as foreign cuisines become more widely shared, so do their once-unusual ingredients. Or maybe it's just because they taste so good.

As the demand for mushrooms grows, so has the number of mushroom cultivators. More and more mushroom farms and gardens are being planted around the world. And with that growth comes new ways of producing mushrooms that are bigger, more flavorful, and faster growing—methods that are making it simpler to grow them yourself.

Why go to the trouble when they're so easily available these days? If you're any kind of a gardener, you already know the satisfaction of eating what you've grown. Regardless of how hard it is, seeing the bounty of your toil makes the work worth it. In fact, growing a mushroom is no more difficult than growing a tomato. You just need to learn a different set of skills.

Mushroom growing is for you if you love to cook and want to jazz up your recipes with unique flavors (without spending a fortune in the produce section). Mushrooms are an excellent source of the fifth flavor known as umami. Umami has been called savory: that dark, deep, satisfying taste that's different from sweet or sour, salty or bitter. It's hard to describe, but if you think of aged cheese, seared meat, and, yes, mushrooms, you've got it. There's definitely a profit to be made in large-scale mushroom cultivation, if that's where your interest lies. What you'll need more than anything else is patience. You can grow mushrooms to enrich your soil, to speed composting, and even to suppress weeds. Or you can cultivate them for medicinal use, making them into teas and tinctures.

The fascination of watching this unique form of life grow may be the biggest draw for a new cultivator. Treat a dead log or pile of straw just right, and in no time, pin-size 'shrooms will start poking up. From the first emerging signs of life to the fully formed final product, the life cycle of the mushroom is nothing short of magical.

1

The Basics

So you've decided to grow your own mushrooms. Now let's figure out how to go about it. You have some decisions to make. What to grow is undoubtedly the first one. Where and how to grow are important, too.

Mushrooms don't exactly grow on thin air. In fact, most of their growing takes place underground, or in a log, or even under a pile of straw, so successfully cultivating them is quite a bit different from growing geraniums or cauliflower. There's a lot to learn.

Understanding how mushrooms grow is crucial to successfully cultivating them. In this chapter, we'll teach you the fundamental facts: What a mushroom actually is. What it likes to eat, and where it can find it. Then, to help you decide which to start with, we'll look at several different mushroom varieties, and give you lots of information to consider in making your choice. Most important, we'll show you how you can successfully replicate in your own home what nature gives the mushroom to survive in the wild.

What Is a Mushroom?

Let's begin with a close look at this unusual entity. A mushroom is actually a fruit—the fruit of a fungus. And the fungus is a network of fibers, called mycelium, which kind of looks like cotton candy.

The fungus can't make its own food, like plants do; it needs to break down the organic matter of a host to survive. And it needs water. So this network of fibers spreads—sometimes for miles—to find a food and water source, and, under the right conditions, matures and produces fruit—the mushroom! Which, when it matures, will start the process all over again.

It all starts with a spore, a single reproductive cell. Mature mushrooms release millions of spores, most commonly from between the gills, which hang under their caps. The spores drift on air currents, and land back on the earth or on plants. Spores are floating everywhere—we breathe them in and out every day!

There are three basic ways mushrooms grow, each defined by how they get nutrition (see "The Three Ways Mushrooms Grow," at right). The saprobic is by far the easiest to cultivate. For the rest of the book, when we talk about growing mushrooms, it's this kind we'll be referring to.

THE THREE WAYS MUSHROOMS GROW

Parasitic/Pathogenic

Food source: another living thing, usually a plant.

How they eat: by stealing nutrition from their host, and killing its cells.

Can they be cultivated? Yes, but it's not commonly done. The host can take many years to grow.

Symbiotic/Mycorrhizal

Food source: another living thing, usually a tree.

How they eat: by a mutually beneficial bond with their host. The fungus offers water, carbon dioxide, and nutrients; the tree gives up oxygen and sugars.

Can they be cultivated? Not very easily, and not in a cost-effective way.

Saprobic

Food source: dead matter, such as wood, leaf litter, dead animal tissue, and manure.

How they eat: by breaking down dead organic matter for nutrition.

Can they be cultivated? Yes, yes, and yes!!

Mycelium

DID YOU KNOW?

- A giant mushroom growing in Oregon is thought to be the world's oldest living organism. It covers 3.4 square miles (8.8 km^2) and is at least 2,400 years old.
- White button mushrooms were originally brown. In 1926, a Pennsylvania farmer came upon a white mutation, which people preferred, and that's why we've mostly eaten white ones for all these years.
- Cremini are just brown versions of white buttons. Portobellos are simply fully mature cremini.
- Mushrooms have been prized since ancient times. The Egyptians thought they could promise immortality, and only royalty could eat them. The Romans called them *cibus diorum*—food of the gods. And in ancient Greece, soldiers ate them for strength in battle.
- Mushrooms are more closely related to humans than to plants, according to DNA sequence analysis.
- There are more than thirty kinds of mushrooms that glow in the dark—great if you're lost in the woods.
- China is said to have cultivated the first mushroom, around 600 C.E. China is still the world's largest producer, supplying more than half of the world's crop.
- Can mushrooms prevent the growth and spread of cancer? Current research says they may, especially the reishi and maitake varieties. Studies are continuing.

Choosing Your Materials

The Mushroom

Now that you know a bit more about mushrooms, the next step will be choosing the kind you'd like to grow. There are several questions to ask yourself. The first should be, of course, which do you like to eat? Mushrooms can vary widely in flavor, texture, and culinary uses.

Next, where will you be growing them? What kind of space do you have available? For a garden setting, ground-dwelling mushrooms like wine caps and blewits can fit in very well. In a wooded setting, choose a mushroom that prefers logs.

Then, there's climate. Shiitakes seem to grow well in most climates; oyster mushrooms are choosier. If you live in a cold climate but have access to a greenhouse, you can also play with agaricus mushrooms, the kind commonly found in the grocery store.

Think about where you will get your growing supplies. How available are they in your area? There may be local garden stores or farms nearby ready to help and, needless to say, buying online is also an option.

What about the cost? That won't matter so much if you're growing on a very small scale. But maybe you'll catch the mushroom-growing bug, and start to think about expanding. The biggest expense is in the growing medium. If you own a forest that needs thinning, cutting your own logs can be a great way to keep the costs down.

Time can be a factor as well. Putting together a wine cap bed can take an hour of your time to produce a summer full of mushrooms. Doing enough shiitake logs to supply the neighborhood will take many days of work in both winter and spring. Choose a mushroom that fits your schedule.

Shiitake

Mushrooms for Home Growing

We'll focus on six mushrooms in the coming pages. Which one should you grow? Here's a look at each to help you decide.

- Shiitake: *Lentinula edodes*. Native to Southeast Asia, where it grows wild on dead branches of the shii tree. Amber to dark brown, with an umbrella-shaped cap. Does well in most climates. Easy to cultivate on logs.
- Oyster: *Pleurotus ostreatus*. Common in Europe and North America. Relatively large, with a creamy to light brown cap. Grows in often stemless clusters on logs, straw, and sawdust. Can be cultivated in containers.
- Wine cap: *Stropharia rugosoannulata*. Native to Europe, but common in North America. Large mushroom with large, dark red cap. Likes to grow on the ground on straw and wood chips.

Hericium

- Hericium: *Hericium americanum*. Also known as bear's head. A compact cluster of long, white spines. Finicky fruiter, but may be grown on logs using the totem method.
- Blewit: *Lepista nuda*. Native to Europe and North America. Lovely lilac/purple hue when young. Grows well on composted leaves.
- Agaricus: *Agaricus*. Many varieties, including the familiar grocery store white button, cremini, and portobello. Grows well on compost, but likes a warm climate and needs a long growing season.

Two others you might want to try:

- Reishi: *Ganoderma lucidum*. Bright red, shelf-shaped with a surface that appears varnished. Used for tea, but never for food. Beautiful, so can be grown just as an ornamental. Grows on logs, but takes its time.
- Nameko: *Pholiota microspora*. Very popular in Japan. Small, amber-colored with a tacky, slightly slimy coating (which disappears when cooked). Can be grown on logs, but by a unique trench method.

Pile of logs to be used as a substrate

The Substrate

What is a substrate? The substrate is the medium in which the mycelium will grow, its food source. Saprobic mushrooms, which we're growing in this book, can be cultivated on various substrates, but it's important to choose the right one. Some fungi are picky about their substrate; others, not so much.

Natural logs, straw, sawdust, wood chips, and compost are common growing mediums; these are the five we'll focus on. Others often used include coffee grounds and manure.

Substrates for Home Growing

- **Logs:** Obtained by cutting living trees into usable sizes and shapes. The logs must then be allowed to rest until all the cells are dead.
- **Straw:** From oats or wheat. It can be chopped to increase surface area, or whole straw can be used. Avoid barley and rye straw because of their antifungal properties.
- **Wood chips and sawdust:** Can be purchased from a local tree service, sawmill, or lumber company. Landscaping chips from a commercial greenhouse cannot be used; they contain dyes and possibly other chemicals.
- **Compost:** From composted table and garden waste. Last year's leaves are a great choice. Make sure you don't include animal bones. How well composted it should be will depend on what mushroom you're trying to grow.

The Spawn

Now that you've chosen which mushroom to grow, and which substrate to use, you need to select the spawn, the material used to inoculate the substrate and start the crop growing. You can grow your own from spores or tissue culture, but it's much less complicated to purchase prepared spawn, a ready-to-use mix of mycelium available in several forms.

Prepared spawn comes in different formats for different applications. If you're growing shiitakes, it's also broken down by when it will work the best in different climates. The strains will each fruit in the spring, but warm-weather spawn is better in summer and cool-weather spawn works into the fall. Wide-range spawn works well everywhere. If you live in a warm climate, the cool-weather strains can even produce over the winter, not a season we usually find mushrooms growing.

There are many reputable spawn suppliers around the world. All of the necessary equipment is available from these suppliers as well. (See the list of suppliers on on page 106.)

Spawn

These are the three formats of prepared spawn.

- **Sawdust spawn:** Finely ground sawdust that is fully inoculated with mycelium. The most common, and usually most economical. Can be used for log or straw cultivation for most mushroom species. Must be sealed in with wax.

- **Dowel spawn:** Wooden pegs inoculated with mycelium. Can be pounded into logs, or driven into a compost pile. Requires more labor than sawdust. Also needs sealing.

- **Grain spawn:** Grain penetrated by mycelium. Works well for mushrooms that don't require soil contact. Must be mixed with straw using a specific ratio for the type of mushroom you're cultivating.

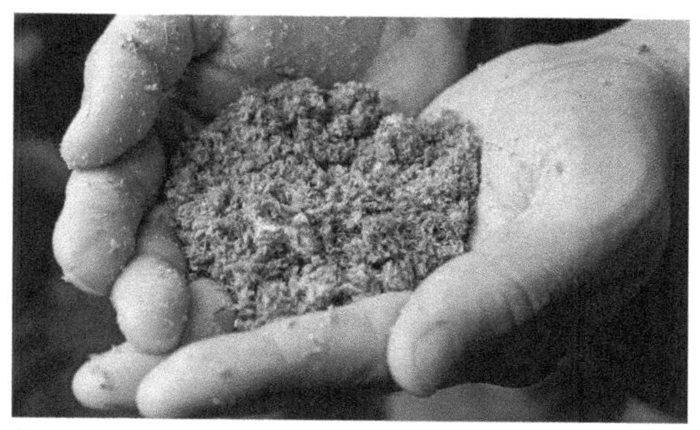

Sawdust spawn

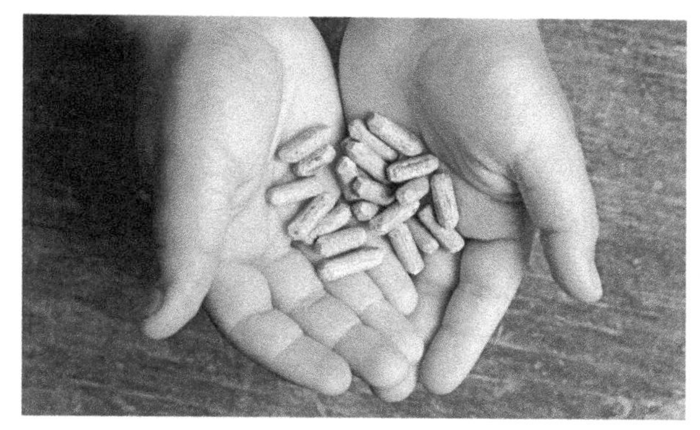

Dowel spawn

Grain spawn

2

Growing Mushrooms on Logs

Take a walk in the woods, especially after a warm rain, and chances are you'll find mushrooms growing. They may be growing at the base of some trees, on a decaying stump, on tree bark, or even on tree trunks.

Many mushrooms like to grow on wood. But you don't need to find a nearby forest to enjoy these delicious fungi. With a little care and knowledge, you can cultivate them in your own backyard, or even inside your house. In this section, you'll find just about everything you need to know for a successful crop, from choosing the right species to selecting the best log, and right up to how and when to harvest.

Right: Shiitakes growing from crib

What to Know about Trees

Yes, many wild mushrooms can be successfully cultivated, but they can be picky. You must match the species of mushroom to the species of tree it's most comfortable with. Some mushrooms grow on a wide variety of hosts; others may be exclusive to a single tree species.

And some trees are out of the question. Aromatic woods—the evergreens such as pine and spruce—must be avoided; they just don't work for any mushroom cultivation. (But they can be useful: they'll help shade your inoculated logs, and can work as a barrier to keep undesirable fungi away.) If you choose the right species, and treat your logs well, a good mushroom log will produce for five to eight years; less desirable trees may only last a year or two.

A forest in need of a trimming is a great place to harvest mushroom logs. Overcrowded trees cause shunted growth; you're making the forest healthier by thinning it. Trees with crowded crowns—the top part of the tree with its leaves and branches—can be thinned as well, though it's best to leave that to the professionals. Again, you're helping the tree: thinning the crown reduces the weight on main limbs, lessening the chance of them splitting off from the trunk.

The trees must be dormant; the best harvesting is done between fall, when the leaves change color, and spring, when the buds open. And they must be alive, to ensure that no competing fungi have started to grow in the wood. Young trees work well, as do branches from older trees.

But how will you know which tree is which? Leaves are an easy way, but we want to harvest when the trees are dormant, and they won't have any! Instead, you'll need to use branching styles, bark patterns, and crown shape, all of which are easy to spot in the dead of winter.

WHICH TREE FOR WHICH MUSHROOM?

Shiitake

All species of oak. Best choice: white oak

All species of maple, especially sugar maple

Birch

Cherry

Hophornbeam

Oyster Mushrooms

Maple, especially box elder

Aspen

Hophornbeam

Red and sugar maples

Reishi

Oak

Sugar maple

Nameko

Cherry

Sugar maple

How to Identify Trees

Branching Patterns

Opposite branching: Look for two branches growing directly across from each other on a larger branch. It can be hard to see on lower branches; check upper ones instead. And do inspect various places on the branches, as smaller twigs often get broken off, giving the illusion of alternate branching. All species of maple have opposite branching.

Alternate branching: In this kind of tree structure, smaller branches alternate which side of a larger branch they grow from. Most hardwoods have alternate branching.

Bark Patterns

Check the texture of the bark. It can be distinctly scaly or grooved, or smooth, flaky, or papery. There may also be unusual knots or splotches of color.

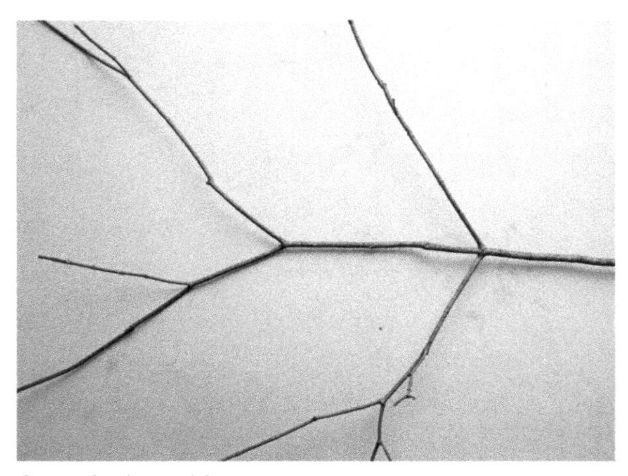

Opposite branching

Alternate branching

Bark of white oak tree

Crown Shape

Trees with vertical branches will have a crown that spreads only near the top, and will look slender. Trees with lots of horizontal branches will have a more spherical crown, and typically a heavier base; it takes a large trunk to support all those large branches.

Now, let's take a look at the trees themselves, and give you a few more clues.

- Oak (*Quercus*) are among the largest hardwood trees. They all have alternate branching. They can be broken into two groups: white, which includes the subdivisions white and bur, and red. All have deeply grooved bark at the base of the trunk, though the limbs can differ, plus large, horizontal branches, giving them a broad, round crown.

  The white oaks have scaly or grooved limbs. They can support mushroom growing for a long time—five to eight years—and offer very high production. Bur oak tends to have thick, cork-like bark that can make drilling deep inoculation sites difficult. The red oaks have smooth limbs. All oaks are excellent for shiitakes, though red oaks tend to fall apart faster than whites, lasting only three to five years.

- Maple (*Acer*) can be broken down into three groups: hard (sugar) maple, soft maple, and box elder. They all feature opposite branching. Maples have heavy horizontal branches, which give them a spherical crown. They can resemble oaks from a distance; look for the opposite branching, which is the key identifying feature.

  Sugar maple has a shorter life than white oak, but produces high yields for three to five years. The trunk bark is heavily scaly; upper limbs are smooth. It often has gray patches resembling blotches of spray paint. Excellent for most mushroom cultivation.

  Red maple and silver maple are the soft maples. Red maple has pale gray, smooth bark; silver maple has extremely scaly bark. Both have a short life after inoculation, just one to two years. Use them only if that's all that's available.

  Box elder usually has a twisted trunk, with tightly grooved bark, and an irregular shape. The bark is greenish. It rots quickly, but it's a good host for oyster mushrooms.

- Hophornbeam (*Ostrya virginiana*), also commonly called ironwood, rarely gets larger than 10 inches (25 cm) in diameter and more typically is only half that size. The wood is very hard and wears out drill bits faster than most other woods. It has a straight trunk with thin, horizontal branches and a spherical but narrow crown. The bark is scaly or flaky at maturity; the smaller branches are smooth. It lasts four to six years, producing well. Good for shiitakes.

- Cherry (*Prunus serotina*) is a common forest tree with black, scaly bark. The smaller branches often have a fungus called black-cherry knot that resembles small, burnt clumps around smaller twigs and occasionally the tree trunk. Cherry has somewhat vertical branches, giving it a spreading, rather V-shaped crown. Great for nameko.

- White birch (*Betula papyrifera*) has distinct, papery bark that often curls as the tree grows. It decays quickly. Birch has horizontal branches and a spherical crown. The bark is tight, and fruitings will occur mostly at the inoculation site. The logs will usually last only two to four years, and it is a low producer. Only use birch if you don't have any other choices.

- Aspen (*Populus*) is easily identifiable by its smooth, pale gray bark, which has a chalky feel. Its wood is soft and the tree doesn't usually grow large. The trunk is straight, the branches horizontal, and the crown spherical. Older trees develop deep grooves in the bark near their base. The wood decays extremely quickly and logs will typically last for only one to three years. Very good for oysters.

Growing the Shiitake Mushroom

Shiitake, oyster, nameko, reishi, and hericium are often grown on logs. Because shiitake is the most commonly cultivated, we'll use it for our first mushroom-growing project. Let's start with white oak, if you can find it; it's shiitake's favorite.

Preparing the Logs

Now that you've chosen your mushroom and tree varieties, it's time to harvest your logs, and then keep them in good condition until they and you are ready to start the inoculation.

First, think about the size. You want to be able to handle them easily. For most people, that means about 40 inches (101 cm) long and 3 to 8 inches (8 to 20 cm) in diameter. When you're ready to stack them, that length will be important. You certainly can harvest larger ones, but they can be heavy and cumbersome. Smaller ones will also work, but they won't last as long. If you're growing mushrooms indoors, however, they can be a good choice.

Once you have the logs, protect them until they're ready to be used. They must be kept moist, and competing fungi and other microorganisms must be kept out. The bark acts as a barrier to both, so handle the logs carefully. If there is visible damage, be sure to wax it shut before the log starts its incubation period. (More later in this chapter on how and when to seal with wax.) Cover them with a simple sheet or tarp, and keep them in a shaded, moist environment. They need to rest for about two weeks. This allows the cells, which are antifungal when alive, to die.

Then, within two months of harvest, it's time to inoculate. Any later may allow other fungi to start colonizing the wood. The exception is if the logs are frozen; they can then be kept for six months as long as they're sheltered from sun and wind.

Log on a worktable

Inoculation

Try to plan your inoculation time for early spring. Fall works, too, but the inoculated logs need six weeks before they can be allowed to freeze.

MATERIALS

Log

Shiitake sawdust spawn

Cheese wax (or other sealing wax)

TOOLS

Safety glasses

Drill with ½ inch (12 mm) drill bit

Inoculating tool (or fingers!)

Metal can

Heat source

Deep-fat fryer or saucepan

Wool dauber or other wax applicator

Apron

Materials gathered for inoculation

Drilling a hole in the log

1. **Drill the holes.** The first step of the inoculation process is providing holes for the spawn. Before you start, put on the safety glasses! Drill holes into the log to a depth of 1 inch (2.5 cm) straight through the bark. Start the first hole about 2 inches (5 cm) from the end of the log. Follow a straight line to the other end, drilling a hole about every 6 inches (15 cm).

   Now rotate the log so your next row is 2 inches (5 cm) lower than the first. Start this row 5 inches (13 cm) in, and follow another straight line, again drilling every 6 inches (15 cm). This will put your holes centered between the holes of the first row, but 2 inches (5 cm) lower.

   Continue alternating rows 1 and 2, and repeat the pattern around the log. You should end up with a diamond pattern.

   You can also space the holes closer together, if you'd like, and increase the number of inoculation sites in each log. This will make the spawn run faster and colonize the log earlier. This is a particularly good idea if you're using trees with a thick bark, because you won't be able to get the holes all that deep. You'll be giving the spawn more access to digestible wood, thus speeding colonization.

Diamond pattern of drilled holes

Spawn slug pushed from inoculation tool

Inoculating logs

2. **Fill the holes.** Now we're going to add the mushroom spawn. This should be done immediately after drilling to prevent the holes from drying out. You can use your fingers, but if you're doing a lot of logs, an inoculation tool can help immensely. It's basically a syringe that will push a slug of spawn tightly into each hole.

   Here's how to use it: Loosen the spawn into a metal can, and then force the inoculating tool into the sawdust with two or three quick stabs. This will fill the end of the tool with a slug of spawn. Then hold the tool directly over the hole and strike it with your hand, or push hard with your thumb, to force the spawn into the hole.

   Either way, whether with the tool or your fingers, you want to make sure the holes are filled to the top and packed tightly. No air spaces! Give each hole a little push with your fingers to check. And rotate the log when you think you're done to make sure you haven't missed any.

Drilled holes filled before sealing

Using a wool dauber to seal holes shut

3. **Seal the holes.** Once the log is inoculated, it must be sealed shut with wax. First, melt the wax; an old deep-fat fryer or saucepan on a hot plate is a good container for this. If you use cheese wax, which we recommend, heat it to about 350°F (177°C). (Paraffin may be used in hot-weather climates; it tends to crack and break off in cold weather. Beeswax is easiest to apply, but it will melt off in hot weather.)

   Be careful! Wear safety glasses and an apron. The wax will be extremely hot. Keep it away from both your skin and your clothes.

   Wool daubers, used to apply stain to leather, make a great applicator. They're often found in craft shops. A paintbrush or sponge will also work.

   Make sure the wax seals the holes completely; any air that gets in can dry out the wood and allow invading fungi in. Any other injuries to the bark should be sealed, too, including knots where branches were removed or broken off, chainsaw marks, and any damage from log-to-log contact. But leave the ends open; this is your best path for water to get in and feed the mycelium. You want to make sure the fungus has the moisture it needs to grow.

Caring for the Bed

Now comes the most critical part of the process: incubation, when the mycelium colonizes the log. It can take between six and eighteen months to complete. During that time, the logs need to be kept out of direct sunlight and have plenty of fresh air and water. Total darkness isn't necessary; any shaded area will work fine. The temperature shouldn't exceed 80°F (27°C), or the mycelium may be damaged or killed. Higher temperatures can also encourage other fungi you don't want to grow.

Lay the logs on rails or blocks above the ground to give them air space; this will also keep any invading soil fungi out. They need water every seven days, so if it hasn't rained for a week, you need to do the watering. A sprinkler or soaking hose is best; it will deliver ample water without drowning the logs or shocking the inner tissues with a temperature change.

Logs stacked in cribs to prepare for fruiting

Stacking

Getting close to harvest now! But first, let's stack up the logs to make that a lot easier. You want this, the fruiting position, to allow you to reach all the mushrooms that have popped up. There are two common methods: the crib method and the lean-to method.

Logs stacked as a lean-to for fruiting

Crib method: Stack two logs on blocks or on a pallet about 24 inches (61 cm) apart. Don't let them touch the ground. Alternate the direction of the logs as they are layered up to the desired height. In cold climates, the cribs should not be stacked too high; you don't want them exposed to the freezing winter wind. Snow can help this, but in years when there isn't enough to cover the logs, use burlap or canvas to prevent moisture loss.

Lean-to method: Lean the logs against a fence or tree at a 45-degree angle, and leave ample space around each one so you can reach the mushrooms when they fruit. If possible, don't allow them to touch the ground to keep any invading fungus out. This method takes up more space, but is just as effective as the crib method.

Harvesting

Once the log is fully colonized, the mushrooms will start to grow. They'll start as small pins, and then, in just a few days, they'll mature to full size. If it's cold out, they may take a little longer.

How to tell when they're ready for your sharp harvesting knife? They'll let you know: just look for something we call veil break. The young shiitakes have a cottony membrane, called a partial veil, that covers the gills and extends from the stem to the edge of the cap. It stretches as the mushroom matures and draws in water, finally breaking when the shiitake has gotten as big as it's going to get. Now is the time to harvest! Once the veil is fully broken, the mushroom is going to flatten out, start losing flavor, and get tougher. It won't hurt to harvest too early, but if you wait, you'll get a bigger mushroom. With a little practice, you'll be able to tell whether the veil has broken; gently slip your finger under the cap, and see if you can feel the now-exposed gills.

Get your knife ready, and cut the mature mushrooms as close to the log as you can without injuring the bark. Because you'll most likely get several more harvests from that same log, you'll want to keep from opening a pathway for invading fungi to contaminate it. And make sure you're only harvesting the kind of mushroom you planted—be wary of other species that may also have grown on the log. The older the logs are, the more likely it is that you'll run into this.

Then, use or refrigerate the mushrooms soon after you harvest. Not sure what to do with them? Later in this book we offer some great recipes.

The logs will rest for a few weeks, and then fruit again. In fact, you may still be getting mushrooms from them as many as six times a year for several years. With each season, there will be fewer and smaller mushrooms. Eventually, the mycelium will have eaten up all the nutrition in that log; all that's left will be a soft, spongy mulch. Time to start some new logs!

Cutting a mushroom away from the log

Growing the Hericium Mushroom

The totem (stacked) method is a great way to use logs of larger diameters. Instead of drilling holes, we'll spread a layer of spawn between lengths of logs and stack them vertically, one on top of the other. The key to success is keeping your totem wet and preventing overheating, regardless of which species you decide to grow this way.

This method works well for shiitake, but it's even more effective for oyster mushrooms and the hericium species. Oysters will colonize many substrates easily—we'll try them on straw in the next chapter—and, if they have the chance to fruit around a stacked log, you may well get large clusters encircling its entire circumference. Hericium are choosier, but let's give them a try because they're so delicious!

Inoculation

MATERIALS

Large log

Large clear or white garbage bag (not black—it might overheat)

A tie for the bag

Hericium sawdust spawn

TOOLS

Safety glasses

Saw

Axe

1. **Set up the totem.** Choose a shaded area for your log. Wearing safety glasses, cut the log to about 2 feet (61 cm) long. Then chop it in half horizontally. Set one section of the log inside the bag. If one side is larger than the other, use the fatter side here.

Cutting logs for a totem

Layering spawn on the bottom log

Stacking the totem

2. **Add the spawn.** Roll the bag to the bottom of the log for now. Place an even layer of spawn, about ¾ inch (2 cm) thick, on the upper face of the log. Push the remaining section of log onto the first. Make sure the top section is well balanced and won't tip off. Wiggle it a bit to settle it into place if it doesn't feel stable.

3. **Enclose the totem.** When the upper section feels secure, pull the garbage bag up around the whole stack and tie it loosely at the top. Don't seal the bag completely, as you'll want to leave a vent for heat to get out and air to get in.

Pulling the plastic bag over the totem

Caring for the Bed

Now let the column incubate for about four months. Occasional watering can help the process, but it isn't necessary, because the bag will retain a lot of moisture.

Harvesting

Look for mushrooms to first appear around the seam of the two sections, and later, anywhere on the log.

How to Force Fruit for a Quicker Harvest

Mushroom fruiting times can be unpredictable. It's hard to know exactly when you'll have enough for that great recipe you've been dying to try. But there is one way, with one certain kind of mushroom, that can give you some control. It's called forcing; you may even have tried something similar with spring bulbs when winter has dragged on a little too long. The mushrooms can be put on a schedule to force them to fruit; the logs are shocked to stimulate growth. But it can be done only with shiitake mushrooms, and even then, only with the warm-weather strains (that is, those that would do well in hot summer temperatures).

Tank with log soaking

Other species of mushroom don't react well to force fruiting; the odds are the fungus will be killed rather than stimulated. Other strains of shiitake don't work well either. Forcing them tends to stunt the future mushrooms or even kill the mycelium—obviously not the result you're looking for.

A log that has been inoculated for at least one year can be forced. First, let it starve for water for two to three weeks. This will make the mycelium branch out in search of moisture, creating a network that will produce mushrooms quickly when water is eventually added. Then, submerge the log in cold water overnight. The rapid drop in temperature and the sudden abundance of water will now trigger the log to produce mushrooms.

Force fruiting does drastically shorten the life of your shiitake logs; it promotes more and faster fungal decay. You will get roughly the same number of mushrooms per log, but it will be over a shorter period of time.

Other Species of Mushrooms on Natural Logs

Although they are less commonly cultivated, two other kinds of mushrooms, nameko and reishi, also can be grown on logs.

Nameko is a flavorful mushroom commonly used in miso, the Japanese soup. It grows well on sugar maple and cherry logs, but it's incubated differently from shiitake. A nameko log does best with direct ground contact. Some people even bury the log horizontally, leaving only the top half exposed. Other than how you place the log, inoculation and incubation are the same as for growing shiitake.

To try it, first choose a shaded area. Then dig a trench half the diameter of the log you're planning to use. Now drill holes and inoculate the log, just as you did for shiitake. You'll want to make holes around the entire log even though only the top will be exposed. This will help the mycelium colonize the whole log, so there's no room for anything else to grow in there.

Seal the log, and lay it flat on the bottom of the trench. Pull dirt back into the trench to cover the bottom half of the log. It's now going to stay this way for the rest of its productive life. Because it's in direct contact with the ground, this nameko-inoculated log will need less watering than one used for shiitake: only about once a month, and if it rains, maybe even less than that.

Nameko only fruit in the very late fall when the temperatures are cool, usually lightly at six months, more heavily at eighteen. Night temperatures below freezing can help trigger this mushroom to fruit. The developing mushrooms will be covered in a gelatinous coating. This can be a turnoff; it keeps many people from growing them. In fact, the coating cooks off quickly, and nameko is an excellent edible mushroom.

Reishi is known for its use in teas and tinctures, and may even give a mild energy boost like caffeine. It's not a mushroom for slicing and cooking. Reishi grows well on sugar maple as well as oak logs, and is treated the same as shiitake. But it does require a longer growing season. If you live in a cooler climate, you may have to grow them indoors or at least bring them in to finish the season.

Bright red, reishi should be harvested after it develops its red color, but before its pores start to turn brown. Reishi can even be grown as a garden ornamental in warmer climates.

Nameko mushrooms

3

Growing Mushrooms on Straw

If you don't have a forest in your backyard to provide you with logs, or even a place to let a lot of mushroom-inoculated logs grow, straw might be the answer.

Not all mushrooms will grow on straw, but the ones that do will grow well. Oyster mushrooms and wine caps are both fond of straw. We'll show you how to grow both, and the best ways to do everything possible to ensure success, from where and how to choose the straw, to how to prepare it and, of course, how to inspire a healthy crop of mushrooms to grow on it.

Yes, there's some prep work involved, but it might be easier and more convenient than chopping down a tree! This method is especially easy if you live near a farmer who grows grain and is happy to supply you with a nice weed-free crop.

Straw stems

What to Know about Straw

Let's start with a definition: straw is the hollow stems of grains left behind after a harvest. While it can come from many different grains, the straw from oats and wheat is what we want for mushroom cultivation. Straw from barley and rye doesn't work as well because they both have antifungal properties that will inhibit mushroom growth.

Never substitute hay for straw! Be sure of what you're getting when you buy that bale. Hay composts far too quickly, and the mushrooms won't have a chance to grow.

We want straw that has as few weeds left in it as possible. The purer the straw, the better the substrate you'll have for mushroom production. A farmer advertising 97 percent weed-free straw is offering your ideal medium. Weeds not only hinder colonization, but also often contain seeds that will try to grow. They'll rot quickly, contaminate the rest of the straw, and ruin your substrate.

The less green you see in the straw, the better. Check the product before you buy it—it will save headaches down the road. If you do find some weeds, be sure to remove them, even though it's a tedious task. Left in, they'll only harm your results.

Three Methods for Preparing Straw

There are three basic ways to get straw ready for cultivating mushrooms. Each has its advantages, depending on which mushroom you're hoping to grow and how you plan to grow it. Before starting any of them, however, you might think of shredding the straw. It's not necessary, but it does give the mycelium a lot more surface area to bond to the straw and a stronger start to their life colonizing it. You can use a leaf shredder if you have one handy.

Straw being submerged in a tank

Method 1: Soaking

How to do it: Straw that will be used to grow mushrooms on the ground needs to be fermented under water for three to five days. Submerge it completely so that no stems are exposed to air. This will drown any seeds that may be in there. The soaking will also ferment the straw and soften its tissues, making it easier for the mycelium to get a good start.

It will have a funky odor when it comes out of the water, and will smell sweet and earthy. You may want to wear rubber gloves: the smell can be hard to get off your hands.

Use it for: wine caps

Method 2: Cold Pasteurization

How to do it: Cold pasteurization is similar to soaking; the difference is adding pickling lime to increase the pH of the substrate.

For every 5 gallons (19 L) of water, you'll need about 13 ounces (368 g) of lime. This will raise the pH to around 10. Oyster mushrooms can tolerate this higher pH, but competing fungi will not. This will help isolate the straw to keep the mycelium network pure.

The straw then needs to drip-dry for 24 hours before you inoculate it. Let it dry outside if you have the space; sunlight and wind will speed the process. But don't let the straw get rained on—this may change its pH. Be sure to cover it if rain is in the forecast.

Use it for: oyster mushrooms

Stirring lime and water

Method 3: Hot Pasteurization

How to do it: Heating is the most effective way to get rid of invading fungi and bacteria before inoculation. You may want to wear protective clothing—a Tyvek or painter's suit, plastic gloves, dust mask, and hair net or baseball cap—to keep both you and the area clean.

Put the straw and water into the container to be heated. Using a gas burner or an electric appliance, heat the water to 180°F (82°C). Hold this temperature for 1 hour. This will kill the competitive bacteria and fungi, but leave the helpful bacteria alive. They'll do a good job protecting the straw from other invaders. If you let the water boil (212°F, or 100°C), you'll kill all the microorganisms, good and bad.

Remove the straw from the water and place it on a metal screen or wire-mesh hardware cloth to drain. Be careful not to burn yourself with that very hot water! If you're doing this indoors, be sure to put a container under the straw to catch the drips. After it's heated, the straw needs to drain in a clean environment for 24 hours. Enough water will remain to feed the mycelium, but there won't be enough to pool and thus provide a breeding ground for the competition.

Use it for: preferred method for oyster mushrooms

Growing the Oyster Mushroom

There are two basic methods for inoculating straw. With the first, the straw is kept in a container; in the second, it's right on the ground. Oyster mushrooms don't need any contact with soil; they grow well in a plastic sleeve or some kind of container. Wine caps and some other mushrooms need that soil contact. They're typically grown outdoors or in a greenhouse. We'll tell you about both. First, oysters.

Growing Oyster Mushrooms on Straw in a Container

Using straw for a substrate to grow oyster mushrooms is quick and effective. The mycelium colonize it far faster than other substrates and may even produce mushrooms as quickly as two weeks after colonization. You'll get high yields from little material, at a minimal cost. Plus, the mushrooms will be cleaner than if grown on wood and will need little preparation after harvest.

Mature oyster sleeve

You can grow oyster mushrooms in different kinds of containers. You just need something that will hold the substrate together, be it the straw we're using here, or as we'll see later, sawdust. The container needs to have openings for the air to get in and the new mushrooms to get out. We recommend plastic sleeves, which are available from substrate suppliers, but almost anything can be made to work.

Have any empty plastic pails or flowerpots around the house? Just give them a good wash, drill some holes in them, and fill with oyster spawn and substrate. Give the mycelium a chance to grow, and soon creamy white clusters will pop out of those holes. Want a great conversation piece? Hang an oyster-growing flowerpot in the garden!

Inoculation

MATERIALS

Small square bale of straw (wheat or oats), 20 to 40 pounds (9 to 18 kg)

Oyster mushroom grain spawn

Plastic sleeves or other container with holes for substrate-spawn mix

Cable ties if using plastic sleeves; a lid or aluminum foil and a rubber band if using containers

TOOLS

Chipper or shredder (optional)

Implements for pasteurizing the straw

Scale

Container to mix the spawn and substrate

Knife or drill to pierce holes in the incubation container

Draining the straw outdoors

Weighing the spawn

Shredding the straw

1. **Prepare the straw.** It's a good idea, but not necessary, to start by running the straw through a chipper or leaf shredder. It will both increase the surface area available to the spawn and make the straw easier to handle. Just remember to remove any weeds from the straw before shredding to prevent decay and likely contamination.
2. **Soak the straw.** Now choose the best method to soak the straw. Method 1 will work, but method 2, cold pasteurization, will work much better. Method 3 is preferred, but not crucial, and it's a bit more work. So let's go with method 2. First, soak the straw from three to five days. Then let it drain for 24 hours.
3. **Measure the spawn and substrate.** The inoculation rate of oyster mushrooms on straw is 5 to 10 percent. This means for every 10 pounds (4.5 kg) of damp straw, add 8 ounces to 1 pound (225 to 450 g) of spawn to the mix. The higher the percentage of spawn to straw, the faster the mycelium will colonize the substrate, and the faster you'll see mushrooms. This can also help prevent invading fungi from getting a good foothold in the substrate. The advantage of using less spawn is you can inoculate more substrate and produce more mushrooms in the long run. Looking for the best of both worlds? Try a rate of 8 percent (13 ounces, or 365 g).

4. **Mixing the spawn.** Start by washing your hands thoroughly. Spores or bacteria on your hands can colonize the straw and outrun the mycelium. Then, roll the bag of grain spawn in your hands to break any clumps into individual seeds. Now sprinkle the spawn over your measured substrate in an even layer. Roll and fold the straw by sliding your hands under it, palms up. Lift your arms up and turn your palms down, folding the straw over itself. Repeat this as least three times. This will ensure that the spawn has made as much contact with the straw as possible.

   You may want to use only a fraction of the spawn each time and do several applications. After each one, be sure to roll and fold the straw again. The individual seeds may fall through the straw and collect underneath it. Roll these seeds back into the straw as you go. When all the spawn has been mixed in evenly, the prepared substrate is ready to be put into containers.

5. **Pack the containers.** Begin with a thin layer of prepared substrate on the bottom of whatever container you're using. Pack it in, making sure to fill all corners. You want to eliminate as much air as possible from inside the vessel so that fruiting happens only at the holes you've created. With the plastic sleeves, pack both corners first as tightly as possible.

   It can be hard to get a good solid mass at the bottom of the bags. Adding small amounts to start will make this easier. It's not a fast process, and you shouldn't try to make it into one. Take your time. Continue adding substrate and packing it tightly as you go. You can use your knee to press down on the plastic sleeves. A bucket or flowerpot can be packed with your fist. Fill the container as much as you can.

   For a solid container, you'll need to add a lid. If you don't have one, aluminum foil and a rubber band works well. Use cable ties for the plastic sleeves and tighten them as much as possible. If you're using a plastic sleeve, pierce it with a knife to make holes about every 6 inches (15 cm) all around the sides of it. These are the escape holes for the mushrooms to fruit.

Mixing the straw and the spawn

Pressing and packing the straw into sleeves

Caring for the Bed

The oyster mushroom mycelium will incubate over the next few weeks. Place the containers in a humid environment, where they'll get at least some light. If you're trying to grow one of the more colorful oyster strains, allow even more light. The colonization will take anywhere from three to ten weeks.

Humidity and temperature both affect incubation times. A room with 85 percent humidity is ideal. As for temperature, each strain will grow best at a different one, so be sure to know which oyster you're growing and try to keep them at that temperature. If you're growing them outside, choose a strain that fits the climate you live in.

You'll see the straw start to change color in just a couple of weeks. It's a sign the mycelia are busy colonizing. Soon after their work is complete, young mushrooms, known as pins, will begin peeking through the holes you've made.

Piercing holes in the plastic substrate to allow mushrooms to emerge

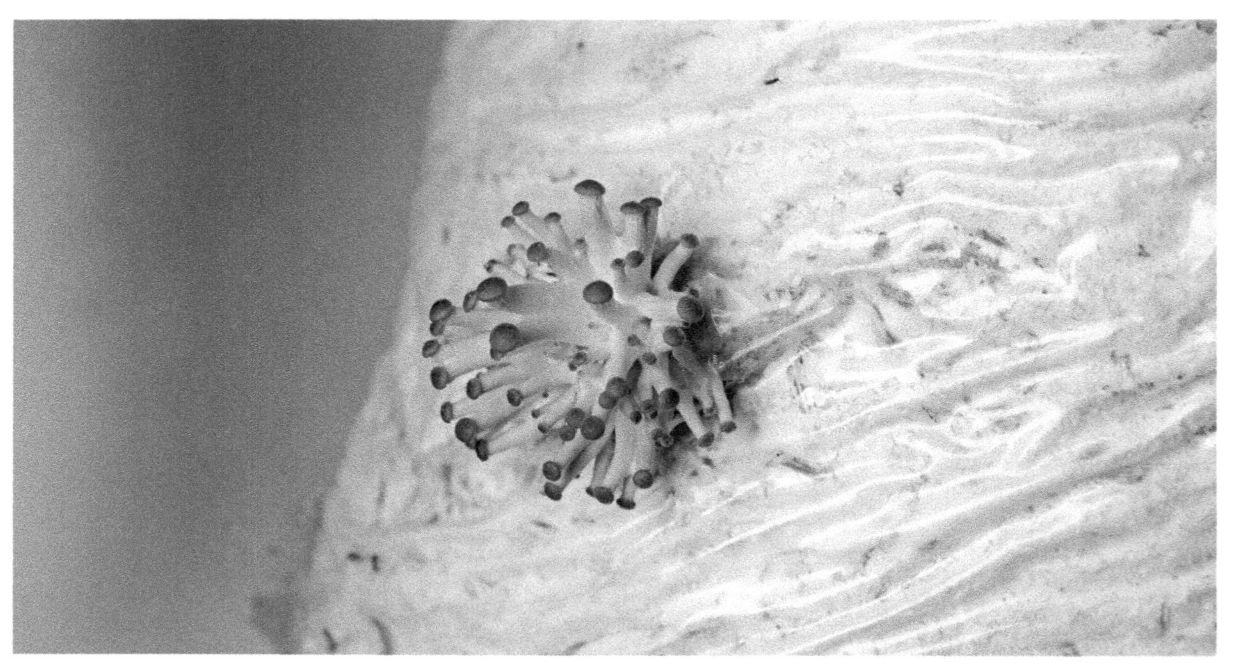

Emerging mushroom pins

Harvesting

Oyster pins take only a day or two to grow to maturity. They develop quickly and should be harvested immediately. Otherwise, they'll rapidly rot—not a good thing if you want to get more mushrooms from this same container.

The mushrooms are easy to harvest: just grab them by the base and twist them off. There will be straw debris connected to the base; trim it with a knife or pick it off with your fingers.

The entire oyster mushroom is tender and edible, so don't trim off more than you need to. The stems have a slightly tougher texture, but are still tasty. Refrigerate the mushrooms immediately after harvest unless you're going to prepare them right away. They have a tendency to decay quickly if not cooled down. Even in the refrigerator, they'll last for only about a week. Find a good recipe posthaste, and cook them up!

Mature mushrooms from straw

Growing the Wine Cap Mushroom

Wine caps are a great mushroom to grow right on the ground. No container is necessary for this one, and there is little chance of failure. The process is simple and quick. Wine caps grow even in the summer heat. Large and flavorful, they'll fill your basket in no time!

Wine caps growing on a bale of straw on the ground

Materials for inoculating straw

Inoculation

MATERIALS

Small square bale of wheat or oat straw; 20 to 40 pounds (9 to 18 kg) (this will be divided in half to make two large beds or one long row)

Wine cap grain spawn

TOOLS

Spray hose or a watering can

Rake

Container in which to soak the straw

Weights

Clear plastic sheeting

Tearing the bale in half

1. **Prepare the area.** With a half bale of straw, you'll be able to build a bed of about 50 square feet (4.6 m2). Choose a place that will be shaded for most of the day. Long periods of direct sunlight will overheat or dry out the beds and hurt the fungus, or maybe even kill it.

   Water the ground thoroughly, and rake up any green matter so that all debris in the bed is dead. Leaving twigs and dead leaves is okay; the wine cap mycelium will colonize those as well and use them for added nutrition. Raking to bare soil isn't necessary, but it will give the bed a cleaner look.

2. **Soak the straw.** We're using method 1 here. Tear the bale in half and put it in a large container. Fill the container with water so that all the straw is covered. Then add weights to make sure it stays submerged. Soak the straw for three to five days so it has a chance to fully break down.

3. **Add the first layer.** Remove some of the straw from its container and let the excess water drain off for a few minutes. It's not necessary to drain it completely. Then shake the straw onto the bed so that you end up with a layer about 4 inches (10 cm) thick. It should be rather fluffy as it piles up. Break apart any large clumps to make it as even as possible. This will both look more appealing and help the spawn run faster.

4. **Apply the spawn.** The next step is to add a layer of spawn. Crumble the spawn between your hands, and sprinkle it over the straw. You should use about 2½ pounds (1.14 kg) of spawn on this layer. Spread

Fluffing straw and making the first layer

Sprinkling spawn on straw

it as evenly and as finely as possible, then break up any clumps that may fall into the straw. The finer the sawdust is sprinkled, the more contact there'll be between the spawn and the substrate. More contact means more thorough colonization.

5. **Add the second layer.** Now we're ready to make the second layer of straw. Remove another batch of straw from the water and let it drain for a few minutes. Shake it over the entire bed just as we did in the first layer and spread it evenly to a depth of about 2 inches (5 cm). Remember to break up any large clumps.

   Then use the back of a lawn rake to tamp the layers together. This will increase contact between the spawn and the straw and help the colonization. Any spawn that doesn't contact straw will eventually die, slowing down the growth of the whole layer. Now apply another layer of spawn exactly the way you did the first one. Once again, use about 2½ pounds (1.14 kg) of spawn. Crumble it, sprinkle it, and spread it evenly across the entire surface. The amounts of spawn are approximate and more can certainly be added for an even faster spawn run.

6. **Add the third layer.** The bed is ready now for the final layer of straw. Remove the straw from the container, but this time, don't drain it. This extra moisture will filter through the bed and keep the core of the layers damp. This layer should be about 1 inch (2.5 cm) thick and even. Make sure no sawdust is visible on the surface. If there is any, cover it with small handfuls of straw to guarantee the spawn is making contact from every possible angle.

   Repeat the tamping process with the back of a lawn rake until the bed looks as appealing as possible. After all, it's going to stay this way for a month or more! Then water the top of the bed with about 1 gallon (3.6 L) of water from a spray hose or a watering can.

7. **Incubation time.** After the bed is nice and wet, it needs to be covered to keep it that way. Cut out a clear plastic sheet and place it over the entire bed and then weigh down the edges. Don't use black plastic; it can overheat and kill the fungus underneath. Leave the plastic in place for four weeks: this is the incubation period.

Adding a second layer of straw

Tamping down the third layer

Covering the bed with a plastic sheet

Caring for the Bed

Inspect the bed about once a week to make sure it's staying wet. You should see water droplets under the plastic. If it seems to be drying out, add 2 gallons (7.2 L) of water. But don't add too much: you'll drown the mycelium.

After four weeks, take the plastic sheet off. Immediately water the bed with about 5 gallons (4.5 L) of cool water. Then peel back the straw in a few places to check for the spawn run. It will look like a white, stringy mat, and have a sweet smell. Success? Replace the straw as close as you can to where it was, but don't put the plastic back on.

Continue adding 5 gallons (4.5 L) of water a week. The mushrooms should start showing up three to five weeks later. If there's no spawn run, or it's weak, keep the plastic on for another week or two.

Lifting the plastic sheet and seeing the spawn run underneath

Harvesting

Wine cap mushrooms are easy to identify; you won't have to worry much about picking the wrong kind. Other mushrooms may grow in the beds, but it will be obvious which ones you're looking for.

It's also simple to remove them: just pull them up from the ground. There will be heavy fibers of mycelium at the base of the stem; they can be long and tough. Cut them and the base off with a sharp knife.

Brush off any loose dirt and debris, then put your harvest in a basket or mesh bag. Trim the bases clean, and put them in the fridge soon afterward. They'll keep for about a week. Unrefrigerated, they can spoil in a matter of hours. After all your work, you wouldn't want to lose them now, would you?

Scooping out mushrooms from the straw

The first mushrooms you'll see will be on the outer edges of the bed, but there are others hiding from you deep inside. They can be hard to find! Look for irregular bulges in the straw. Touch them with the palm of your hand, and feel for a hard spot. That's the mushroom's cap pushing up the straw. Then try tapping the beds gently with your palm; you may find even more hiding in there! Carefully scoop out the straw to expose them, gently remove them, and then replace the straw.

4

Growing Mushrooms on Sawdust and Wood Chips

We've already seen how to use two different mediums—one from a forest, the other from a farm—to cultivate a successful mushroom crop. Next on the list is something you may have in your own home, if you're into carving or carpentry, or know someone who is. We're talking now about sawdust and wood chips. They make excellent substrates for certain kinds of mushrooms. Wine caps love wood chips; in fact, they often grow without any help at all—whether they're wanted or not—wherever a pile of chips or mulch can be found. Oysters are equally happy; sawdust closely mimics what they munch on in the wild.

If it works in nature, the odds are good that both these mushrooms will also do well in your own backyard—with a little help, of course, which you'll find in this chapter.

What to Know about Sawdust and Wood Chips

So where can you get these substrates? Sawdust collected at home or bought at a local sawmill, or wood chips from a tree service, are ideal, as long as they're free of chemicals. Commercial wood chips and mulch should not be used; they often contain fungicides and other chemicals. Avoid cedar chips and dust as well. Cedar is resistant to fungal growth and will make it hard for the mycelium to get a strong start. It will likely die before being able to do any colonizing.

Fine sawdust

Growing Oyster Mushrooms on Sawdust

Cultivating oyster mushrooms on sawdust is not all that different from growing them on straw. Why should it be? They'll need nutrition and water, humidity, and the right temperature, regardless of what they're growing in.

As with straw, you will need a container to hold the sawdust and mycelium. You can use many different types of containers, from a flowerpot to a plastic bucket to a metal coffee can. Anything that will contain the sawdust without letting it fall out will work fine. And as you may remember from the previous chapter, you'll have to punch some holes in it. You'll mix the spawn at a particular ratio to the sawdust, always with clean hands. You'll pack the medium tightly into whatever container you choose.

Oyster mushrooms naturally grow from wood; they're often found on dead or injured trees. So sawdust—which of course is nothing but tiny particles of wood—is a great medium for growing them. For most strains, you'll want to avoid sawdust from spruce, pine, and any of the other aromatic woods. Sawdust from aspen, maple, and box elder works well.

Materials for inoculation

Inoculation

MATERIALS

4 pounds (1.8 kg) fresh sawdust

1-gallon (3.6 L) container

Oyster mushroom sawdust spawn

TOOLS

Vessel in which to soak sawdust

Screen to drain sawdust

Drill or knife to pierce holes in container

Scale

1. **Soak the sawdust.** Soak the sawdust for 24 hours to soften the tissues and give the mushroom an easier chance to get an immediate foothold after inoculation. Make sure all of it is under water. After 24 hours, remove the sawdust, drain it on a screen, and let it drip-dry for another 24 hours. As was true with growing on straw, too much water can cause problems at inoculation, so make sure it drips long enough.
2. **Prepare the container.** Scrub whatever container you've chosen with soap and water. It needs to be clean before you can inoculate it. Then pierce holes in the container

(continued)

There are many different species and strains of oyster mushroom, and they'll each act differently at different temperatures and moisture levels. The strain you choose will determine how you treat the spawn to get it to fruit most effectively. Some want lots of light; others need less. Some demand a warm climate; others will do fine in cooler places. Your spawn supplier will give you the particular details for their product, but the overall process is the same for all of them.

Soaking the sawdust under water

Scrubbing container and piercing holes

so oxygen can enter and the mushrooms can grow out. Make them about ⅜ inch (1 cm) wide, and space them about 6 inches (15 cm) apart.

3. **Mix the spawn.** If you've already tried growing oysters on straw, you'll know the spawn and sawdust need to be mixed in a specific ratio. It's the same here: the spawn should be 5 to 10 percent of the weight of the sawdust. So if you're using 4 pounds (1.8 kg) of sawdust, mix in 3.2 to 6.4 ounces (90 to 180 g) of spawn. And make sure you have clean hands when you do the mixing so you don't introduce any other fungi by accident.

4. **Pack the container.** Once the spawn and sawdust are evenly mixed, you're ready to fill your container. Pack it as tightly as possible so the main source of air is through the holes you pierced earlier. Then put a lid on the container; if you don't have one, aluminum foil or plastic wrap can work. Whatever you use, it should be airtight.

Mixing the spawn and sawdust

Packing the sawdust into a container

Caring for the Bed

Oyster mushrooms take more time to colonize on sawdust than they do on straw. But as with straw, they need humidity. Levels above 80 percent are ideal. If that's not possible, be sure to mist the holes, and possibly even put a plastic sheet over the container to keep the humidity in. Incubation may take as long as three months. Higher humidity can speed this up.

Misting the container

Harvesting

Once the containers are colonized, you'll start to see pins coming out from the holes you drilled. Humidity is crucial now. If the air is too dry, the mushrooms will shrivel and die before they reach any usable size.

As they get bigger, you'll see the oysters developing multiple caps. They typically grow in clusters and may have a dozen or more lobes in each cluster. When the mushrooms grow to the size you want, it's time to harvest. It's remarkably easy: simply grab the cluster toward the base and twist gently. The mycelium will yield its grasp and the mushrooms will be free. Oyster mushrooms should be refrigerated soon after harvest unless they are to be used immediately. They have a shelf life of about a week in a cooler or refrigerator.

Oyster pins coming out of the container

Growing Wine Caps on Wood Chips

Wood chips are a great substrate for growing wine caps. These mushrooms are natural composters that feed on the cellulose in the wood; it offers a stable nutrition source for the mycelium as they grow. They will grow more slowly than they would on straw. But wood chips offer at least one advantage: they'll be a much more attractive addition to your garden.

You'll want the chips freshly cut. Wood chips from more than six months ago may already be hosting other fungi; these can keep your mycelium from getting a good foothold. Better to give your wine caps a fighting chance! Also avoid chips that have been treated with fungicides, dyes, or other growth-inhibiting chemicals. These may kill the mycelium before it even starts to grow.

Begin by clearing the ground of any large debris to give the bed an even contour. This is partly cosmetic, but may also help you get a consistent colony of mycelium throughout the entire bed.

Inoculation

MATERIALS

Fresh wood chips

Sawdust spawn for wine caps

TOOLS

Rake

Spray hose or watering can

Bucket or other vessel for soaking wood chips

Screen to drain wood chips

Scale

Clear plastic sheeting

Weights

1. **Prepare the bed.** Decide what size area you want to cover. Choose a shaded space, as we have before. You'll need 1 to 4 pounds (0.45 to 1.8 kg) of wood chips for each square foot (929 cm2) of bed. The thicker the bed, the longer it will take to colonize. Now use a rake to clear away any large rocks or sticks, unless they're to be part of the final design. Water the ground heavily so the bottom layer of chips won't dry out too much.
2. **Soak the wood chips.** Soaking your chips for 24 hours before inoculation will do two things. It will soften the wood, giving the wine cap mycelium an easier path in, and it will give the fungus a burst of water to get started. Place the wood chips in a bucket and soak thoroughly with water.
3. **Add the first layer.** Drain the chips on a screen for at least 8 hours. Then spread them on the prepared area in a thin, even layer just deep enough so you can't see the ground underneath. It should be about 1 inch (2.5 cm) thick. Unless you find it unattractive, you can leave in any organic debris that's mixed in with the wood chips.
4. **Add the spawn.** The bed is now ready for a layer of spawn. You will need approximately 4 ounces (113 g) of sawdust spawn for every square foot (929 cm^2) of area covered in wood chips. Grind the spawn between your hands and sprinkle it on the bed evenly, covering each piece of wood with spawn.

   You can now add several more layers. How many depends on how deep you want the bed to be, and how fast you want it colonized. The more layers, the slower the colonization. Begin each new layer with 1 inch (2.5 cm) of chips, sprinkle on the spawn, and water it lightly to give the mycelium the moisture it needs to expand. But don't overwater and drown the developing fungus. This should sound like a familiar refrain by now!
5. **Add the final layer.** The last layer of the bed is the one that will be visible. Spread a layer of wood chips over the existing layer, or layers, to a depth that will cover all of the spawn from sight. Water this top layer generously, cover it with a plastic sheet, and weight the sheet down to prevent it from blowing away. We need it held in place to maintain the humidity inside. Leave the sheet in place for 28 days as the bed incubates.

Watering cleared space for bed

Mature flush of wine caps

Caring for the Bed

Now remove the sheet and check for spawn run: thick, white strands of mycelium running through the layers of wood chips. It will have a pleasant smell, and may even reach beyond the edges of the bed. If the spawn run looks weak or discolored, water it again and re-cover with plastic for another two weeks. Once the plastic is off, wine caps should start to develop in six to eight weeks.

Harvesting

Wine caps grown on wood chips are easy to harvest. Just pull the mushroom from the ground and trim the base with a knife. Immature mushrooms will be the most flavorful and have the best texture. Don't worry if they look paler than the usual deep burgundy, even nectarine-colored. It's just because you grew them in the sun. Same flavor, though they may not keep for as long as those grown in the cool shade.

If you're growing wine caps to decorate a rock garden or other landscaped area, you might want to let them get big and beautiful. Wine caps are sometimes referred to as garden giants; they can grow to enormous size—as large as 1 foot (30 cm) across, and weigh more than 1 pound (450 g)! No need to remove them until they weaken and fall over. But you can't eat them at this point; only fresh, firm mushrooms should be considered edible.

Harvesting wine caps from wood chips

5

Growing Mushrooms on Compost

Need a little help with your composting? Could your garden soil use a bit of beefing up? Mushrooms to the rescue! Several species are adept at breaking down garden and table waste and turning it into usable, nourishing soil.

Blewits are the star here; the previous year's leaves are their preferred food. They're not fussy; they like the leftovers from your table and from your garden, too. They're easy to cultivate; the coming pages will show you how.

We've spent a lot of time on a wide variety of seemingly exotic mushrooms, the kind that didn't enter your culinary lexicon until recently. We've found they're not so hard to grow after all. The common button mushroom might seem lowly at this point, but let's end with a look at that species, agaricus; we'll find they're kind of finicky and maybe one of the more difficult to grow.

You're a mushroom-growing expert now, so why not give it a shot? In this chapter, we'll add these last two to round out your repertoire.

What to Know about Compost

Composted garden and table waste or manure compost can make excellent substrates. Growing mushrooms on waste materials will break down the organic matter into usable soil and is an easy way to speed the composting process.

For simple compost cultivation, blewits are a great mushroom choice. They spend the summer digesting the organic matter, and fruit in the cooler weeks of late fall. Leaves raked up from the previous year work particularly well; garden and table waste is also a good choice. The fungus will break it down, and then, in the following year, you'll be able to till it all back into the garden to give your soil a great nutritional boost.

Agaricus is another great mushroom for compost cultivation, but it requires a longer growing season. A greenhouse can help by adding some warmth to the end of the year if you live in a cooler climate. People in warmer climates will have a long enough growing season so they won't have that trouble.

Growing Blewits on Leaf Litter

Blewits grow naturally on leaf litter on the forest floor. They're typically a fall mushroom, fruiting very late in the year. Starting a bed of blewits in leaf litter is an easy way to get edible mushrooms after much of the mushroom season is over. They grow right on the ground on decayed matter.

Blewits can appear in large groupings—sometimes by the hundreds!—and are a favorite among experienced mushroom hunters. Why hunt them down when you can easily grow them yourself? If you have a pile of leaves from last year at your disposal, you have a great substrate for blewit cultivation, and you won't have to go ramping through the woods to find your next meal!

Inoculation

MATERIALS

Last year's leaf litter

Blewit sawdust spawn

TOOLS

Rake

Watering can

Clear or white plastic sheeting

Weights

Materials for inoculation

Watering the area for the new bed

1. **Prepare the bed.** Very little preparation is needed to make a blewit bed. Midsummer is the best time to start. Find a shaded area of whatever size you'd like. Keep in mind that a single 5½ pound (2.5 kg) bag of spawn can cover at least 50 square feet (4.6 m^2). Rake debris to expose the soil surface. A heavy watering at this point will give you a nice source of humidity as the mycelium develops. Water until the ground is fully soaked.
2. **Add the first layer.** Using the previous fall's leaf litter, cover the prepared area to a thickness of about 2 inches (5 cm). Blewits are aggressive, so sterilizing the substrate isn't necessary, and they'll make short work of any bacteria left in the leaves. Any leaves will work, but if you use pine needles, mix them with the leaves of other trees.

(continued)

3. **Add the spawn.** Add one bag of blewit spawn to the leaf litter. Stir it around, breaking up all the chunks. Try to make as much contact between the spawn and substrate as possible. The more contact, the faster the spawn will colonize the leaves.
4. **Add the second layer.** Cover the spawn layer with another layer of leaves. This one should be about as thick as the first, but it can be thicker if you'd like. Gently stir this layer into the layer beneath it. This will ensure the top layer gets colonized as well.
5. **Incubation time.** Give the bed a good watering. Don't oversoak it and drown the mycelium, but give it enough water so that the leaves stick together. Then cover the bed with a plastic sheet, weighting it down, and leave it there for about two months. Once again, clear or white plastic is preferred.

Adding spawn to the leaf litter

Covering the bed with plastic sheeting

Blewits starting to fruit

Caring for the Bed

Peel the plastic back and water about once a week, replacing the sheet afterward. Keep watering until the season is over, or until the beds have been exhausted. As the nights get cooler in the fall, remove the plastic completely and give the bed a good watering. This will trigger the mushrooms to fruit.

Harvesting

The blewits will appear in large troops and grow quickly. A good size of blewit is about 4 inches (10 cm) across. Pull them from the ground and trim the base to remove the debris. Indications they're ready: the gills are crowded and attached to the stem with a small notch, the base of the stem is often swollen and with lots of visible mycelium and leaf litter attached, and they have a rubbery feel.

Blewits can be tricky for beginners to identify. Although they're typically lilac to purple, as they age they are much paler and can even be faintly brown. And though they smell faintly of orange juice, not everyone detects this odor in the same way. In fact, odor can be a confusing tool for mushroom identification. But there are other mushrooms that can look similar, so making a spore print may be a good idea if you're not familiar with blewits. A good regional field guide can be a handy tool to help you safely identify them. (More on spore prints and guides in the next chapter.)

Growing *Agaricus* on Compost

Mushrooms found in grocery stores, such as white button, portobello, and cremini, are from the genus *Agaricus*. They can be cultivated at home, but they are more finicky than other mushrooms. The large commercial operations use sterilized manure to grow them, but they can be grown on compost as well.

You'll want to use compost that's well decomposed. *Agaricus* isn't a good mushroom for breaking down fresh garden waste; it likes material that's had a chance to decay for at least a year. Many of the *Agaricus* species are temperature sensitive and won't do well in cooler climates. A greenhouse will be an immense help by adding a few weeks to the end of the growing season. Inoculation should be done in late spring because the mycelium take a full summer to colonize.

Raking the ground

Inoculation

MATERIALS

Well-decayed garden and table waste

Agaricus sawdust spawn

TOOLS

Rake

Watering can

Clear or white plastic sheeting (optional)

Weights (optional)

Making a layer of well-composted material

1. **Prepare the bed.** Rake the ground clean in an area about 50 square feet (5 m^2). A sunny area is a great choice, because this mushroom needs heat to grow. Water the soil to help keep moisture levels up and speed the spawn run.
2. **Add the first layer.** Make a layer of well-composted material about 2 inches (5 cm) thick. Avoid meat and bone scraps. *Agaricus* wouldn't digest this in nature and it can be a breeding ground for bacteria.
3. **Add the spawn.** Spread one full bag—about 5½ pounds (2.5 kg)—of sawdust spawn over the top of the compost. Work it into the compost layer with your hands, breaking up all of the clumps as you go. Make the sawdust particles as small as you can so you've covered the most surface area possible. Work the spawn and substrate until they're well blended. You can add another layer of compost on top to help keep the bottom layer damp.

Caring for the Bed

Water the bed heavily about once per week. Covering the bed with a plastic sheet can help keep more moisture in place and help with spawn run, but it's not required. If you do use it, avoid black plastic.

The mushrooms will fruit in late fall or even winter in warmer climates. In cooler climates, they may not fruit at all if left outside. This is where a greenhouse can be helpful. The warmer parts of the world are the natural range for many *Agaricus* species. There are cooler-weather species, but they tend to be difficult to grow and aren't as appealing. *Agaricus subrufescens* is a great one to cultivate. It has a strong odor and flavor of almond extract and is a delicious edible mushroom.

Harvesting

Pull the *Agaricus* from the ground and trim the base of the stem. They will look similar to the grocery store varieties. If you've cultivated *Agaricus subrufescens*, they'll have that distinct odor that other mushrooms in the species lack.

HOW TO TELL *AGARICUS* FROM *AMANITA*

There are many species of *Agaricus* and they all look fairly similar, but beginners may have trouble distinguishing them from mushrooms in the genus *Amanita*, which includes some that are dangerous, so be careful in identifying them. Luckily, you can tell the two apart by noting some salient features.

- *Agaricus* has gills that are grayish to pinkish while developing and mature gills that are dark brown or even black. *Agaricus* also produces a dark brown spore print.
- *Amanita* has gills that are white and will produce a pure white spore print.
- The two species may look similar in stature, and both may have a ring or skirt on the upper stem, known as an annulus. Pay attention to gill color, and if that's in question, do a spore print. It's the most accurate way to distinguish the mushrooms from each genus. (We'll explain how in the next chapter.)

An uncovered compost heap in a backyard

Turning the compost heap

How to Make Your Own Compost Pile

A compost pile can be a great help for growing mushrooms or, really, for any kind of vegetable or flower. It's not hard to do; you just need a designated space (as small as a few feet, or as big as you have room for), a source of water (rain will do if you don't live in a dry region), and of course, some stuff to put in it. Got yard waste? Table scraps? Coffee grounds? Then you're at least halfway there.

First, allocate a space. You can make a pile somewhere outside, buy a compost bin, or build your own enclosure. There's a lot of information out there online and in bookstores on how to construct one or where to buy one. You can also track down a local expert at your nearest garden center; they may even sell enclosed bins.

Well-balanced compost will contain a blend of animal manure, decayed green matter, and composted fiber, such as straw. You need to alternate the layers: a layer of brown material, like the straw; a layer of green matter, such as garden waste; and a layer of aged animal manure.

Manure that is at least a year old is especially good for mushroom growing. Manure that has aged even longer, for three or more years, will make excellent soil and is high in the kind of nutrition mushrooms need. However, don't ever add animal tissue or bones to your compost pile.

Garden waste can just be heaped up and left to compost naturally, but turning it over once a week will speed the process and help aerate the final product. Tear the plants apart into the smallest pieces possible; this will help them break down faster.

You can also add sand to your compost; sand will help it better hold in nutrition. It won't hold water in well, but sand does do a good job with organic matter. Newspaper can be another good addition, especially if you add red worms to the shredded paper. Shred the paper as fine as you can before adding the worms. This will make an especially good top layer for a compost heap or a compost box. Keep the newspaper moist and in the dark; too much moisture loss will kill the worms.

If you're interested in growing wine cap mushrooms, you can use them to turn straw into compost for the cultivation of other mushrooms. Wine caps can reduce 12 inches (30.5 cm) of straw into ⅛ inch (3 mm) of black soil in just one year. Wood compost or composted sawdust is great if you're opting to grow blewits. Neither, however, will work for *Agaricus*; as we've said previously, they demand compost that's well decayed.

One last important tip: always wash your hands after working with compost; it contains lots of bacteria. Wearing rubber gloves is recommended.

6

Problems and Solutions

The decision to grow your own crop of edible mushrooms is not one to be taken lightly. It involves a lot of time and effort and, as is the case with many new endeavors, there are certainly things that can go wrong. So though we've tried to give easy-to-follow, understandable instructions, you may feel a bit daunted by this new undertaking.

Maybe it just seems overwhelming; you're worried about making some bad mistakes right off the bat. Perhaps you're further down the road and have a harvest, but you wonder whether what you've got is what you should eat. Did some not-so-edibles mix in with the mushrooms you thought you were growing? Or perhaps you're sure you did everything right, but somehow that mushroom bed looks wrong.

This chapter will help calm those jitters. We'll tell you how to start slow and build your confidence, and if problems come up, we'll show you how—whenever possible—to correct them.

How to Start Slowly: Using a Kit

You'd like to grow mushrooms, but it seems a little intimidating. You're not quite ready to make the leap into logs and straw and sawdust. The problem is, you're a bit nervous about making the investment, in both hard work and hard cash, when you fear it might turn into a fiasco.

The solution: mushroom-growing kits. They're a great choice for beginners and an easy way to learn the basics. Plus, they're easy on the wallet if you do decide this isn't for you. Kits are readily available worldwide from spawn suppliers, usually costing $12 to $40. There are more expensive options, but for a beginner, there's no reason to spend more. Each one will contain a preformed mixture of mushroom spawn, substrate, and water. It will look like a large loaf of bread. You won't need any special tools to get it going, and it won't require much attention, either.

There are dozens of species of mushrooms that can be grown with these ready-made kits, and, in fact, many mushrooms do better in them than in natural substrates such as logs or straw. The kits won't produce as many mushrooms, but there will be enough for a few meals. The mushrooms will typically be clean and easy to harvest.

Another advantage is the time they save. When your kit is delivered, it's just about ready to start making mushrooms. There is no inoculation process; that's done by the kit's manufacturer. There's also no colonization period; the kits are incubated before they are shipped. In many cases, the kits are ready to fruit upon arrival. Prepared kits also take up little space. They can even be used as a decorative centerpiece on a kitchen table or they can be placed out of sight on the back of a countertop.

Prepared mushroom blocks come with a set of instructions. Each manufacturer will have slightly different guidelines, but they will all follow the basic plan described here.

MATERIALS

Ready-made mushroom grow block (included in the kit)

2-gallon (7.2 L) or larger container

Large plastic bottle with cap

Shallow tray or cake pan

4 wooden skewers (often included in the kit)

Sheet of plastic cut to approximately 36 x 36 inches (90 x 90 cm)

Spray mist bottle

All materials needed to use the kit

A block with bulges

Inserting skewers, then covering the block with plastic sheeting

1. **Remove and inspect the kit.** Open the kit and check the mycelium colonization. If the exterior of the block is pure white with protruding bulges over the surface, it's ready to fruit and you can skip to step 3. If the block is mostly brownish with maybe some white areas, it needs to be stimulated, a simple process of giving the mycelium a large dose of water.
2. **Soak the prepared block.** Put the block in a container large enough so that it can be completely covered in water. Add the water. The block will be incredibly buoyant and you'll need to add weight to keep it fully submerged. Here's where the plastic bottle comes in. Fill it with water and place it right onto the spawn block. Make sure the whole block is under water and none of it is exposed to air. If any part dries out, it will draw water away from the rest of the block when it's removed, and potentially harm the mycelium.

   Leave the block in the water for 8 to 12 hours, to make sure it's completely soaked. Then take it out and let it drip for an hour to remove excess moisture on the surface.

(continued)

3. **Prepare the block for fruiting.** Place the mushroom block on a tray or in a cake pan. Any flat surface with sidewalls will work; you're just trying to catch extra water that could make a mess. Then spear a skewer into the block at each corner at an angle just shy of vertical; these will keep the plastic sheet in place. Being just out of vertical will let the plastic hang without touching the spawn block.

   Drape the sheet over the skewers so it touches the tray's surface; the idea is to block any air currents that would dry out the block. Then put the block somewhere it will get light, but not direct sunlight, which can dry it out and even overheat the mycelium. Either of these can kill the mycelium, and there'll be no mushrooms for you.

4. **Fruiting.** Some blocks will produce mushroom pins within a day or two, while others may take as long as two weeks. Peel back the plastic and give the block a healthy misting of water at least twice a day. If you can, try misting four to six times a day; you'll get even better results. Spray the entire block until water beads up on the surface and runs into the tray, then replace the plastic. As mushrooms start to form, continue to mist, but just once a day. The mushrooms will take three to five days to reach maturity.

5. **Harvest and rest.** Harvest the mushrooms by cutting them with a sharp knife as close to the block as you can. Any extra tissue should be removed to prevent bacterial growth on the block. After the mushrooms are harvested, the block will need to rest. This will allow the mycelium to regroup from the work of fruiting mushrooms. Remove the plastic sheet from the block and let it to dry out for one week.

Trying to fruit a block immediately after harvest can put a lot of stress on the mycelium. The result may be stunted mushrooms or even no more mushrooms. After the seven-day resting period, the block can be brought back into fruiting by starting at step 2.

You'll get several harvests before the block is depleted. Each one may be a little smaller, both in mushroom size and in number. This is from the mycelium digesting the substrate. The first fruiting typically produces the largest and the most mushrooms. Each subsequent one will yield fewer mushrooms than the fruiting before it. The mushrooms won't always be smaller, but don't be surprised if they are.

Mushroom pins growing on block

Piles of different mushrooms

How to Identify What You Have: Making Spore Prints

Figuring out what mushrooms you've grown can sometimes be problematic. You know what you've planted, of course, but other fungi may try to horn in on your crop. Although we've provided descriptions and photos of all the mushrooms we recommend growing, there may still be times when you're just not sure. One good way to tell whether what you've just harvested is what you want is by making a spore print. It will ease your worries, and it's easy to do. Fun, too!

How to Do a Spore Print

The color of a mushroom's spores is an excellent way to determine what kind it is. The problem is, mushroom spores are too small to see with the naked eye, and even under a microscope, it's impossible to figure out the color. A print, which piles up millions of spores on top of each other, will make the color jump out.

A perfectly mature mushroom

MATERIALS

Mature mushroom

Paring knife

Aluminum foil or a small pane of glass

Jar or bowl

Mushroom identification guide with photos

1. **Remove the stem.** First make sure you have a perfectly mature mushroom: not too young, but not too old, either. It must be fully developed to get a good print. Younger mushrooms may not have spores that are ripe yet. The spores won't fall and won't have developed enough color. Overly mature mushrooms don't work well, either; they may have already dropped most of their spores. They also can decay quickly, which may leave both water streaks and excess material on the print.

   Once you've found a good candidate, cut its stem as short as you can. You want to be able to lay the cap flat, with its gills facing downward. This will help the spores fall straight down, without having to travel far.

(continued)

Placing mushroom cap with gills down on glass

2. **Place the mushroom.** Lay out a piece of aluminum foil or glass that is at least slightly larger than the mushroom to be printed. These two products work best because they show all colors. If you have a mushroom with pale spores, they won't show up on white paper; the same is true for dark spores on black paper. Put the mushroom gill-side down on the foil.
3. **Cover the mushroom.** Using a jar or bowl, cover the mushroom. This prevents any air currents from blowing the spores away. Spores are designed to float on the gentlest breeze, so even minimal air movement can make a print fail. The idea is to just let gravity do the work.

   Let the mushroom sit undisturbed until the print is ready. Some will produce a visible print in a matter of 2 hours; some may take a full day. A safe bet is to leave the print where it won't be disturbed for about 8 hours.

Mushroom covered with jar on glass

4. **Finish the print.** Carefully remove the cover and gently lift the mushroom cap straight up. The print will be directly beneath where the cap was. Now, examine the color closely. Look it up in your mushroom identification guide. That's all there is to it!

Be creative! You needn't discard your spore prints once you've used them to identify your mushroom. They can be turned into art! With a little practice, you can use the mushrooms' different colors and densities to make some creative pieces. You can even print through stencils. Making them is easy enough for kids to have fun with as well.

Art made by using a variety of different colored spore prints

5. **Use a mushroom guide.** Which guide should you use? There are so many out there! The most important thing is to find a good one for your own region. *Mushrooms of the Midwest*, by Michael Kuo and Andrew S. Methven (University of Illinois Press, 2014), is great for that area, for instance; *Mushroom Picker's Foolproof Field Guide* by Peter Jordan (Anness Publishing, Ltd., 2010) is ideal for Brits. Neither is appropriate if you live in the western United States. Or France!

   When you've got a guide you think will work, check inside. It should have mushroom photos, not drawings or paintings. The colors never seem to be correct in an illustrated book, and the mushrooms are always perfectly shaped. Nature isn't really like that, is it? It makes the books too hard to use.

What Might Go Wrong, and How to Fix It

Now here's the trickiest dilemma: there's clearly something wrong with your beds. Mushrooms can be finicky, and these apparently aren't happy.

To successfully colonize, mycelium needs a steady supply of water and fresh air—but just enough, not too little, and not too much. If they get too dry or run out of fresh air, there's a risk they'll be damaged or killed. The best way to prevent this is to keep up with regular watering and make sure the substrate can breathe. Never completely seal off growing mycelium from air. They don't require a lot, but they do require some.

Overwatering

The potential threat here is mold. If mold starts to form on the substrate, it typically means the environment is too wet. Even though the fungus requires water, too much can cause problems. Be sure to let the substrate go through a drying period between waterings. Once mold appears, it can be difficult to deal with, so let's try to stop it from ever starting.

We only see the fruiting structures of mold; it has mycelium that can go deep into the substrate. Fortunately, mushroom-growing substrates are porous, so you'll just need to treat the outside of the substrate to kill the mold inside. White vinegar or hydrogen peroxide should kill the mold mycelium without doing too much damage to the mushroom mycelium.

Start by carefully pouring one of these liquids into a clean spray mist bottle. Then gently mist the substrate's surface. The mold should die in about an hour. Don't apply more vinegar or hydrogen peroxide for at least 24 hours. It needs to evaporate before you add more to prevent buildup and possible damage to the deeper mushroom mycelium.

If the mold returns, try misting again. This isn't a guaranteed way to eliminate mold recurrences, but it has been proven to work most of the time. If the mold doesn't seem to be affected by the spray, the substrate may be too deeply infected. Then it's time to dispose of the substrate and start over.

Underwatering

This can cause problems, too. People forget or get too busy to keep up with a watering schedule. Missing a few waterings isn't going to kill a fungus, but not watering at all will. The substrate will dry up and die if it's not watered regularly. At first, there's usually enough moisture within to support life even if the surface feels dry, but the mycelium will eventually use all that moisture, too. With most substrates, anything more than a week is too long to go without adding water.

Plastic sheeting does help hold water for longer, and we've recommended it in earlier chapters. There's no substitute for adding fresh water to each bed. When the substrate gets dry, adding water is the only solution. Water it heavily and often for two to three days to try to get back to normal. If the mycelium has already died, it's too late and you'll have to toss the substrate.

In either case, discard any spent or contaminated substrate to prevent future outbreaks of contamination.

Hanging flypaper near a mushroom bed

Insects

Insects and their larvae are another concern in mushroom cultivation. They can be almost impossible to eliminate during spring and summer. They're mostly after mature mushrooms for food and a place to lay eggs.

To keep them away, don't leave any mushrooms on the substrate longer than they need to be. Harvest all mushrooms as soon as you can and leave as little tissue as possible. If flies or beetles do get a good foothold, you'll need to get rid of them to protect the crop. Don't use chemicals; you're growing food, after all, and you don't want to serve them to your guests or yourself.

Flypaper or a flytrap may help reduce the number of insects that come into contact with the mushrooms, but preventing them from arriving is the best method. Protect the substrate as much as possible when in an indoor setting. Keep screens in place, don't leave doors and windows open, and keep the substrate away from other things that may attract insects.

Cleanliness, regular watering, and protection of the developing mycelium are the keys to preventing problems that may put the health of the mycelium in jeopardy. Healthy mycelium produces healthy mushrooms.

7

Processing and Preparation

Congratulations! You've got a successful mushroom harvest on your hands. So successful, in fact, that you're not sure what to do with all of them! Sure, you've been cooking up a storm. You probably won't be able to gobble them all up the day you harvest. We want to make sure they don't go to waste and that you get to savor them for many months to come.

In this chapter, we'll tell you the best way to keep them for both the short term and the long, from a week to a year. Some for now, and some for later: it's up to you. We'll make sure they'll keep for as long as you want them to, and be delicious no matter when you choose to use them. You wouldn't want to waste all the effort you've put in to growing them, would you?

Assorted fresh mushrooms

Storing

Mushrooms will keep well in your refrigerator for a week or longer, if you make sure not to trap moisture in their container. First, don't wash them; all that extra water will hasten their decay. Don't ever store them in plastic bags; there's no way for any moisture to get out. Stiff plastic containers with vent holes are fine, especially if you add a layer of paper towels under the mushrooms.

Paper bags also work well, but if the mushrooms are going to be in the refrigerator for more than a day, the bag should be left open to get any extra humidity out. Small cardboard or fiber containers without tops are great; they allow the mushrooms to breathe and also offer some structural protection.

These will add storage life to your mushrooms, but there may still come a time when a few have been left in the fridge too long. How to tell? Discard any mushrooms that develop odd colors or odors, become wrinkled, or show obvious signs of decay. Don't wait: decay spreads quickly from one mushroom to the next, so you'll probably need to get rid of all the mushrooms in that same container if you don't act quickly. Next time, eat them up faster!

An open box of mushrooms in the refrigerator

Dried mushrooms

Drying

Maybe you can't use all your mushrooms right away, and the neighbors have had their fill, too. What to do? They'll need to be prepared for long-term storage, and drying them is one way to do this.

Drying is the preferred storage method for most mushrooms, though it works better for some varieties than others. Shiitake mushrooms dry extremely well and can keep in a dried state for years. Oyster mushrooms and wine caps dry readily, but the texture of the reconstituted product can be a little too soft for most people. Still, they can be used in soups or other dishes where the mushroom isn't the star ingredient.

An electric dehydrator is the most efficient way to dry your mushrooms, but it isn't necessary. Mushrooms can be air-dried with a fan, sun-dried on screens outside, or even dried inside in the oven.

The basic principle is this: airflow is more important than heat. All mushrooms should be dried at a temperature of 110°F (43°C) or less, if possible. This will ensure the mushrooms dry, not cook. If you cook the water out of the mushroom, it will dry, but the finished product will lose its fresh flavor. Done right, a well-dried mushroom will keep most of the qualities of its fresh counterpart. It can be used in much the same way.

Mushrooms should be dried until they're no longer pliable. "Cracker crisp" is the way to describe the perfectly dried mushroom. If you use an electric appliance, depending on the machine and the size of the load, the mushrooms can be completely dry in 12 to 20 hours. Sun-drying usually takes two to three days. Oven-drying is the fastest—2 to 6 hours—but it yields a lower-quality product.

How to Dry Shiitake Mushrooms

1. **Prepare the mushrooms.** Clean the mushrooms with a dry brush. If any debris remains, you can use a damp towel to gently wipe the outside. Then trim the stems as short as possible.

   They can be dried with the stems in place, but it will affect the way the mushroom loses moisture. The stem's top can hide water and create a pocket that could start the decaying process. Drying them until the stem is dry may dehydrate them too much. It's best to remove the stems and avoid all these issues. You can dry the stems separately, and use them later in stock.

   You can also choose to slice the mushrooms; it can speed drying times and make them easier to work with later. You don't need to, but it is recommended. The thinner the mushrooms are sliced, the more surface area there'll be, and the faster they'll dry.

2. **Arrange on racks.** Whether you're using an electric dehydrator or another source of moving air, you need to arrange the mushrooms so they don't touch each other. Letting air flow around them will help them dry quickly and evenly. You want the process to be as fast as possible so the mushrooms don't have time to decay before they're dry.

   If you're sun-drying them, arrange the mushrooms on a screen or other material that will allow a good deal of airflow. You can also lay them out on newspaper and set up a fan to blow on them. But if you do, you'll need to flip them every few hours to keep moisture from building up on the side touching the paper. Just remember that you want as much surface area exposed to air as possible.

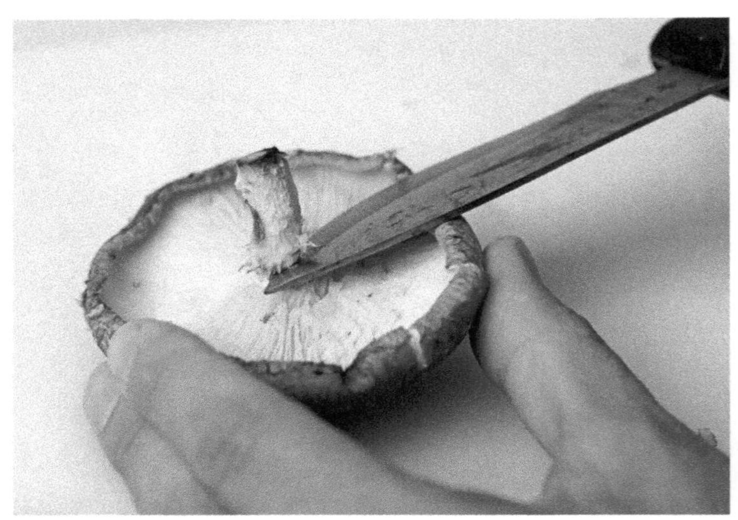

Removing the stems

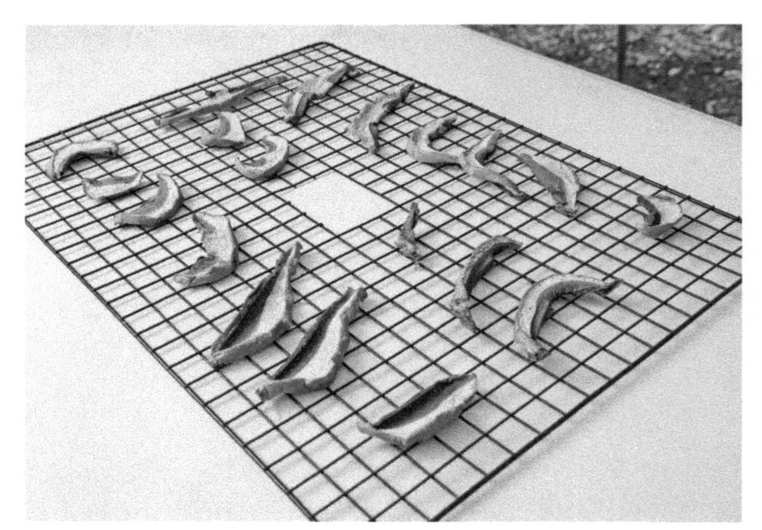

Rack for drying mushrooms in the sun

3. **Dry.** The drying time for shiitake mushrooms can vary greatly for several reasons, with the first being how many mushrooms you're drying at once. A dehydrator will dry one rack of mushrooms quickly, but it will take longer if you're doing multiple racks. Make sure they've been dried to the point of being no longer pliable.

   Another factor is whether they're whole or sliced, and if sliced, how thinly. Whole caps will dry, but it will take longer than if you expose more flesh by slicing. Remember, the more exposed surface area, the better. The temperature you're using will affect how long it takes as well. Higher heat will dry a mushroom faster, but you won't get as good a quality.

   Drying outdoors is a great and inexpensive method, but there are variables here, too, that affect the timing. Relative humidity is one of them; if it's too humid, the mushrooms may not dry at all. Spring and fall are great times of year for outdoor drying; summer, depending on where you live, may prove to be more difficult. Whether it's in direct sunlight or shade matters, too: sun can help a mushroom dry faster, while mushrooms left in the shade tend to lose water more slowly.

   Hanging the mushrooms to air-dry works well and yields a high-quality product. Slice the mushrooms once the long way and make a large necklace using a needle and thread. Cut about 4 feet (1.2 m) of thread and thread a needle with one end. Pierce the needle through the sliced mushrooms until the thread is about half full. Tie the ends of the thread to each other and hang in a sunny area with good airflow. The mushrooms will dry in two to three days.

Using an electric dehydrator

Making a necklace for air-drying

To dry in the oven, slice the mushrooms to a thickness of about ¼ inch (6 mm). Lay them out on a baking sheet and place them on a middle rack in the oven. Set the oven to 150°F (65°C) or even 120°F (49°C) if possible. Leave the oven door partially open and let the mushrooms dry for 4 to 6 hours, flipping once during the cycle.

4. **Store the dried mushrooms.** Once they're dried, the mushrooms should be kept in an airtight container out of direct sunlight. Glass jars, heat-sealed bags, and locking-seam bags work great. They should be kept in a cool, dry place; a kitchen cupboard or pantry is ideal for long-term storage. Squeeze out as much air as you can from the bags before you seal them, but be careful not to damage the mushrooms inside. Jars should be filled as full as possible to limit air space within. You don't have to hide them out of sight: mushroom-filled jars can look quite decorative displayed above kitchen cabinets.
5. **Use them.** Dried mushrooms will need to be rehydrated before you cook with them. The most common way is to soak them in warm water or stock for 5 to 20 minutes. You should keep them soaking until they're completely soft all the way through.

Mushrooms drying on a rack in the oven

Mushrooms in glass wire-top jars

Once the mushrooms are soft, remove them from the soaking liquid, drain them, and give them a gentle squeeze to remove any excess water. Don't throw out that leftover liquid—it's filled with concentrated mushroom flavor. Strain it to get rid of any debris that's fallen to the bottom of the soaking container. Then you can use it to sub for some of whatever liquid is in your recipe. It will add an extra flavor boost to soups, stir-fries, risottos, and many other dishes.

Dried mushrooms also can be ground into powder to be used as a seasoning or soup thickener. A coffee grinder works great for pulverizing them.

Rehydrating dried mushrooms

How to Dry Oyster and Wine Cap Mushrooms

The process for drying oyster mushrooms and wine caps is much the same; the only major difference will be how long they take to dry. Both of these mushrooms dry quickly. They're good candidates for outdoor air-drying, but an electric dehydrator will also work well. The finished product should be dried until it is cracker crisp, just as with shiitake mushrooms.

Rehydrating is much quicker as well. These mushrooms take on moisture a lot faster and it's easy to make them soggy by soaking them too long. Keep an eye on them, as it may take as little as 2 minutes to get them to a usable state.

Pan-frying mushrooms

Freezing

Freezing is another fine way to store any mushroom. There are a few steps you need to follow to make sure you get the high quality you want. All mushrooms to be preserved should be fresh and firm. Cooking them first will ensure they'll have good flavor and texture. You don't need to cook them thoroughly; partial cooking is enough to keep the mushrooms from crumbling in the freezer. The process is the same for all mushrooms you want to freeze.

Freeze mushrooms in containers that hold what you'd normally cook for a meal, rather than in one large container. This will make using them a lot simpler.

Freezing Fresh Mushrooms

1. **Prepare the mushrooms.** First, clean the mushrooms. Brush off any visible debris. If any dirt sticks, use a damp towel to clean the surface. Then slice the mushrooms into strips; this will save you a step after they're frozen and make the process much easier down the road. They don't need much cooking time; blanching or lightly sautéing will do. If you cook them completely, they'll lose their fresh flavor in the freezer. You can season them if you like while they're cooking.
2. **Freeze.** Rather than cramming your mushrooms into a freezer bag and forcing it into the freezer, you'll get a much better result if you first freeze them separately on a baking sheet. This way, they'll hold their shape better when you're ready to use them. Line the sheet with waxed paper or parchment paper and arrange the mushrooms so they don't touch each other. Put this in the freezer and let them freeze solid. Once the mushrooms are frozen, then you can put them in a freezer bag.
3. **Use them.** Ready to use the frozen mushrooms? It's easy, because all the work has already been done. Simply allow the mushrooms to thaw and then warm them back up, or add them to a dish that's already cooking. Frozen mushrooms retain much of their flavor and texture, and can be used just like fresh mushrooms.

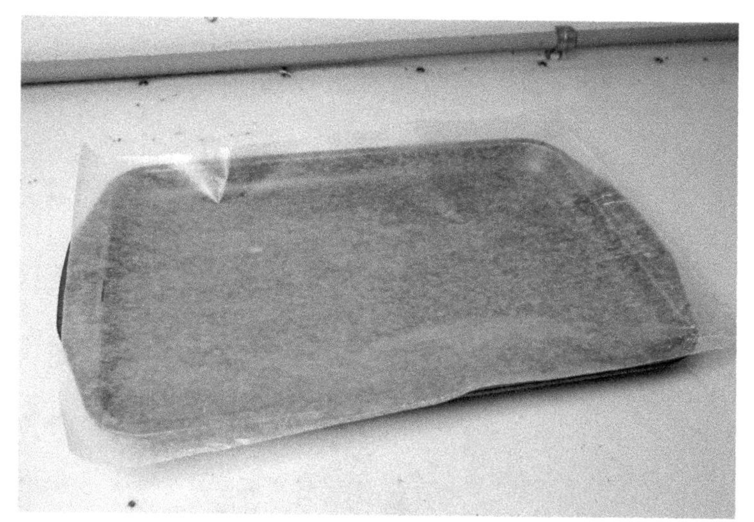
Baking sheet covered with waxed paper

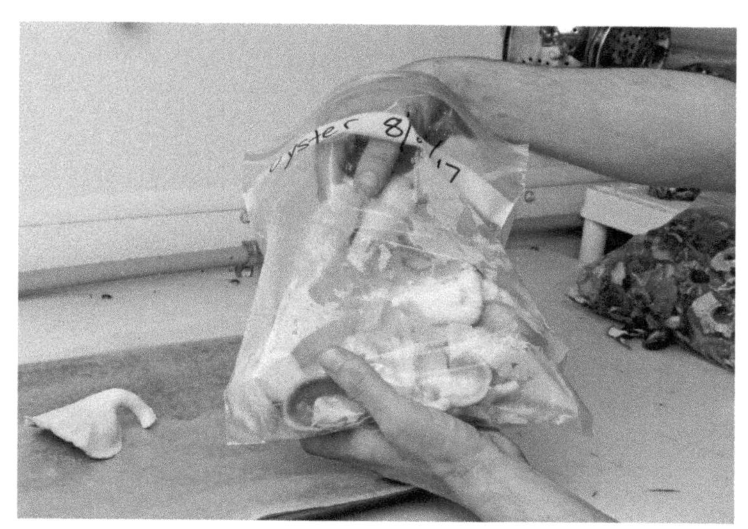
Filling bags with frozen mushrooms

8

The Finished Product

Now comes the reason you've coaxed and coddled all those mycelia for all these many months. You've got mushrooms, and now you get the chance to eat them!

Regardless of the type, mushrooms are incredibly versatile, and there are practically countless ways to prepare them—sautéing, stewing, or braising; making mushroom soup or mushroom sauce; as an appetizer, a main course, or a salad addition (some people even make them into a dessert!).

The easiest choice is simply to sauté them up: just heat some butter (or oil, but mushrooms and butter bring out the best in each other), throw in some slices, maybe add garlic, and cook until they've absorbed all the liquid they've just sweated out. Serve and enjoy. You've just made a great side dish or garnish for almost any entrée.

If you're British, perhaps add them to toast; French, make an omelet. Japanese? Miso soup. Just about every cuisine you can think of uses mushrooms in one tasty way or another.

We'll tell you how to get the most out of your mushrooms in this chapter. Taste buds ready? Let's go!

Different kinds of pans

Cooking with Mushrooms

While popping them in a hot pan is the easiest way to cook most mushrooms, that's not by a long stretch the only thing you can do with them. Mushrooms can play a tasty part in soups, eggs, stews, and many other meal preparations. They marry particularly well with shallots, garlic, cream, and bacon, and, combined with those ingredients, make some delicious dishes.

You'll most probably be cooking mushrooms for their taste and, often, texture. But aside from adding flavor, there is also some nutritional value to most mushrooms. They're low in calories, fat-free (until you sauté them!), low in sodium, and high in some important nutrients. Depending on the type, they can provide protein, thiamin, folate, magnesium, phosphorus, potassium, zinc, copper, manganese, selenium, niacin, riboflavin, calcium, iron, and vitamins B_6, C, and D. Whew—that's a lot to cram into one tiny mushroom. Of course, they each have just a bit of these, but it's nice to know that, in addition to their deliciousness, they're good for you!

Each type of mushroom has its own unique flavor and texture and will absorb flavors differently. One type may work better than another in a particular dish, depending on how they're prepared and what they're paired with. It's also a matter of personal preference. Some people like a nice chewy mushroom in their soup; others will lean toward the kind that almost melt in your mouth.

Experimentation is the best way to learn which mushroom works best with each type of dish you're preparing. We'll give you this guide to start with, then try one of our yummy-sounding recipes, like maybe that Spicy Asian Oyster Mushroom Soup. Will it taste as good or better with a wine cap? There's only one way you'll find out—get your apron on and get cooking!

The Mushrooms and How to Cook with Them

Wine Cap

What they taste like: More like a vegetable than a traditional mushroom. Think asparagus or raw potato.

How best to cook them: Wine caps are extremely versatile. They're especially tasty pickled, or combined with lemon juice or white wine. Red wine can sometimes bury their delicate flavor. Try sautéing them in butter or olive oil; again, watch out you don't overwhelm them. They don't need a long cooking time, and can be sizzled up quickly to preserve their sharp, clean taste.

Shiitake

Oyster

What they taste like: Mild flavor. Fishy, floral, or anise-like. However, taste is not as important as texture with these.

How best to cook them: They're good at absorbing other flavors, so soups are a great choice, as are recipes where they're combined with a lot of other savory ingredients. High heat or a long cooking time can leave a bland mushroom. Oysters can hold their bright colors well and can give great visual appeal to any side dish.

Gray oyster

Shiitake

What they taste like: Strong! Meaty or yeasty, with a chewy texture.

How best to cook them: Not as versatile as some of the milder mushrooms, but definitely have their uses (and quite a few fans!). Very common in

WARNING! READ THIS BEFORE YOU BEGIN

If you're not absolutely sure of what you've harvested, don't eat it!

No home-cultivated mushrooms should be eaten raw. Bacteria may be present on the mushroom tissues that can make you sick.

Asian cuisine, especially in stir-fries. But they also work well with pasta and other Italian dishes. The stems are too tough to eat, but add a lot of flavor to stocks and soups. Just be sure to fish them out before serving. Or puree them—great flavor without the off-putting texture.

Blewit

What they taste like: Slightly fruity or citrusy. Soft texture.

How best to cook them: Just about anywhere cooked mushrooms are a good fit. Be sure to cook them thoroughly; they're mildly toxic when raw. They can turn delightfully crunchy when cooked for longer periods.

Hericium

What they taste like: Similar in taste and texture to boiled lobster.

How best to cook them: Best on their own. Using them in a complex dish is a waste of their unique flavor. Cook them thoroughly; they're bitter when undercooked. Clean out any debris trapped inside them; feel free to tear them apart to get at it.

Nameko

What they taste like: Nameko have a rich flavor, similar to shiitake, but much sweeter. They have a slimy coating on the cap when fresh, but this disappears quickly when they are cooked. Some consider them the best-tasting mushroom, whether wild or cultivated, of all.

How best to cook them: Commonly used in Japanese miso soup. They are also excellent sautéed and should be cooked until crisp for a nice crunch.

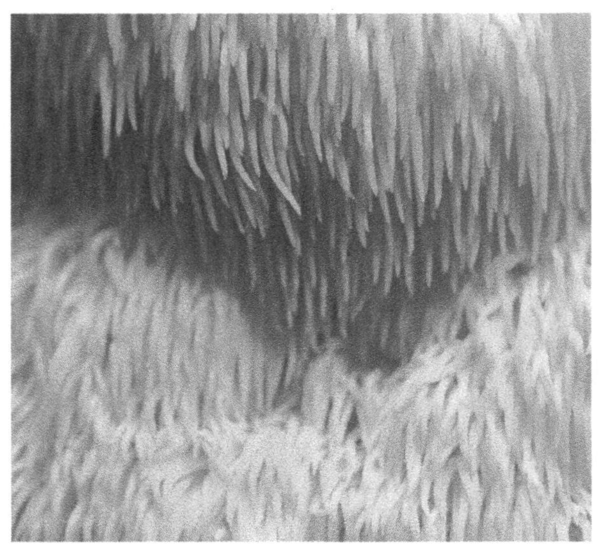

Hericium

Nameko

Agaricus

What they taste like: Usually bland, though some varieties have a taste of almond extract.

How best to cook them: Often sautéed. But they, too, are great at absorbing other flavors, so try them in recipes with strong-flavored ingredients. Home-cultivated agaricus can be used almost anywhere grocery store mushrooms are used, but they offer a much sweeter flavor that doesn't work for every dish.

Tips for the Tastiest Mushroom Meals

- To wash or not to wash? Some chefs say don't, it will make them mushy; just wipe them clean with a damp towel. Others say that a quick wash won't hurt them and will clean them more efficiently. One thing the experts do agree on: don't soak them. That *will* destroy the texture for sure. They'll be waterlogged and won't brown, no matter how hard you try. If you do choose to wash them, do it just before using them. Don't wash and then put them back in the fridge.

- When you're sautéing, don't crowd the pan. The mushrooms will stew, not sear, in all the moisture they release. Leave some space between them. And make sure your pan and fat are both hot, hot, hot! That will help the liquid evaporate, and you'll get some beautiful results.

- Ever heard of sweating mushrooms? It's a common way to prep them, usually sliced, before adding to a stew or a similar dish. Here's how: Heat some oil over low heat in a medium to large skillet, depending on how many mushrooms you're working with. Add the mushrooms, maybe a little salt, even a bit of wine if you like. Then cover the skillet—that and the heat level are the big differences from sautéing—and cook until the mushrooms have released all their juices. It shouldn't take any more than 10 minutes. Now, to get that nice sear, uncover the pan, turn up the heat, let the liquid boil away, and keep cooking until they look deliciously browned.

- If a recipe calls for fresh and all you have is dried, you can substitute 1 pound (450 g) of fresh for 2½ to 3 ounces (71 to 85 g) of dried.

Wiping dirt off mushrooms

Pickled Wine Caps

My father created this dish from an old pickled fish recipe my family has used for years. He made a few adjustments in the amounts, took out the cloves and a lot of the sugar, and we've been savoring pickled mushrooms ever since. Now you can, too. Wine caps that are 1 to 1½ inches (2.5 to 4 cm) will work best. You'd have to slice the larger ones to fit them into the jars, and that won't look nearly as nice when it's done.

7 (1-quart, or 1 L) Mason jars and lids

Water, for processing the jars

60 to 80 wine caps, depending on size

7 cloves of garlic

FOR BRINE

7 cups (1.6 L) water

7 cups (1.6 L) apple cider vinegar

2½ cups (500 g) granulated sugar

2 tablespoons (30 g) coarse salt

2 tablespoons (30 g) pickling spice, with cloves removed

NOTE

You can also use agaricus mushrooms in this recipe.

Place the jars, without lids, on the jar rack of a canning pot, with water that comes up at least 1 inch (2.5 cm) over the top of the jars. Put the pot on the stove, cover it, and turn the heat to high. When the water has boiled, remove the jars. Wash the lids in hot water.

Pack the jars with the mushrooms, leaving ½ inch (1.3 cm) of headspace on top. Add 1 clove of garlic to each jar.

TO MAKE THE BRINE: Mix the brine ingredients in a saucepan and bring to a boil. Pour the boiling liquid over the mushrooms in the jars, again leaving ½ inch (1.3 cm) of headspace.

Cover the jars, place them on the rack in the canner, and boil them for 20 minutes. Carefully remove them from the canning pot.

Let stand for 1 week before opening.

—

Seven 1-quart (1 L) jars

Spicy Asian Oyster Mushroom Soup

This recipe comes from my longtime friend Chad Patrin, executive chef at the 5 O'clock Club in Cumberland, Wisconsin. The restaurant regularly buys my wild mushrooms for its standard menu. A few years back, I had a surplus of oyster mushrooms and called to see whether Chad would take them off my hands. How could he say no? He created this recipe especially for them, and now, whenever I have extra oysters, I know the best place to send them. Take note: When he says "spicy," he means it!

½ cup (80 g) diced onion

¾ cup (90 g) diced celery

¾ cup (100 g) diced carrot

3 tablespoons (45 ml) olive oil

1 pound (454 g) oyster mushrooms, sliced

1 tablespoon (10 g) minced garlic

1 tablespoon (10 g) minced ginger

2 quarts (2 L) chicken or vegetable stock

1½ tablespoons (23 ml) soy sauce

1½ teaspoons (7.5 ml) hot sauce (or to taste)

½ cup (60 g) cornstarch

½ cup (125 ml) cold water

In a large pot, sweat the onion, celery, and carrot in the oil until the onion is translucent, 6 to 8 minutes. Add the oyster mushrooms, garlic, and ginger. Cook until the mushrooms are tender, 10 to 12 minutes. Do not brown.

Add the stock, soy sauce, and hot sauce, and bring to a boil. Mix the cornstarch and water in a cup to make a slurry, and whisk it slowly into the simmering soup. Allow to simmer for a few minutes, until the soup turns glossy and thickens. Serve.

—

8 servings

NOTE

You can also use shiitake mushrooms in this recipe.

Mushroom Miso Soup

Nameko mushrooms cry out for miso soup; they're an essential ingredient in this well-known Japanese dish. The authentic recipe calls for dashi, a stock of seaweed and dried fish, but it takes some time to make, and ingredients that may be hard to find. Here we'll sub water or vegetable broth for a quick, convenient snack instead. Inauthentic, but still tasty.

3 cups (700 ml) water or stock

4 ounces (112 g) nameko mushrooms, sliced

2 tablespoons (34 g) red miso, or more if desired

2 ounces (58 g) silken tofu, cut into ½-inch (1.3 cm) chunks

1 scallion, thinly sliced

Heat the stock in a medium saucepan. Add the mushrooms and simmer until tender, about 5 minutes. Remove ¼ cup (60 ml) hot stock from the saucepan and put in a small bowl. Mix in the miso until smooth and then return to the pan. Turn off the heat. Do not let the soup boil after adding the miso.

Distribute the tofu between two bowls. Pour the soup over it, and top with the sliced scallion.

—

2 servings

NOTE

You can use white miso instead of the red; it will have a milder flavor.

Tia's Mushroom Sauce

Tia Bismonte-Duncan and her husband, Kelly, run the Lairdchurch Bed and Breakfast in the tiny town of McKinley, Wisconsin—population 15!—not far from my mushroom farm. Though they call it a B&B, Tia and Kelly entertain evening guests often. They host adult game nights, have wine and dinner parties, and treat their guests to wonderful meals throughout the year. This recipe, adapted from one Tia learned at Le Cordon Bleu in Paris many years ago, is often on the menu.

FOR ROASTED GARLIC

1 head of garlic

2 tablespoons (30 ml) extra-virgin olive oil

FOR SAUCE

4 tablespoons (60 ml) extra-virgin olive oil, divided

4 tablespoons (60 g) unsalted butter, divided

2 ounces (56 g) pancetta, cubed

2 pounds (908 g) oyster mushrooms

½ of a large Vidalia or other sweet onion, minced

1 tablespoon (3 g) chopped fresh tarragon

Sea salt and freshly cracked pepper, to taste

2 cups (470 ml) heavy cream

¾ teaspoon red pepper flakes

½ cup (50 g) freshly grated Parmesan cheese

TO ROAST THE GARLIC: Preheat the oven to 400°F (200°C or gas mark 6). Peel and discard the papery outer layer of the garlic head, leaving the skins of the individual cloves and keeping the head intact. Using a sharp knife, cut about ¼ inch (6 mm) off the top of the bulb, exposing the interior of each clove. Place in a baking pan with the cut side up. Drizzle the olive oil over the exposed cuts of the garlic bulb. Rub the oil in with your fingers. Cover with aluminum foil and bake for 30 to 35 minutes, or until the head feels soft when pressed. Let it cool slightly and then squeeze the garlic cloves from their skins with your hands.

TO MAKE THE SAUCE: Heat a large frying pan over medium heat. Once the pan is hot, add 2 tablespoons (30 ml) each of the olive oil and butter. When the fat is hot but not smoking, add the pancetta and cook until slightly browned, 3 to 4 minutes. Add the mushrooms and cook until soft, 6 to 8 minutes.

Push the mushrooms and pancetta to the side of the pan. Add the remaining 2 tablespoons each of oil (30 ml) and butter (30 g) to the middle of the pan, then add the onion and cook for about 3 minutes. Add the roasted garlic and tarragon and cook for an additional 2 minutes. Season with salt and pepper.

Add the cream and raise the heat to high, bringing to a boil. Once boiling, reduce the heat and simmer. Add the red pepper flakes and simmer until almost thick enough to coat the back of a spoon. Stir in the Parmesan cheese and serve.

—

4 servings

NOTE

This recipe is great with pasta or over pork tenderloin or chicken thighs.

Risotto with Wild Mushrooms

An Italian classic, and for good reason. Risotto comes from northern Italy, where rice, not pasta, is often the star, and this is one of its most stellar uses! Somehow rice and stock, with just a few additions, turn into something exceptionally creamy and satisfying. Maybe it's flavored with patience; it does require some. Or time; this isn't a five-minute meal. Then there's the stirring and stirring and more stirring: many say that's the real secret to perfect risotto. We're adding shiitake mushrooms to the mix, with their deep, dark flavors. And by using dried as well as fresh, we up the umami even more. Yes, it's time- and labor-intensive, but the reward is oh-so delicious.

1 ounce (28 g) dried shiitake mushrooms

Warm water

6 ounces (170 g) fresh shiitake mushrooms, brushed free of any dirt

5 to 6 cups (1.2 to 1.4 L) chicken, beef, or vegetable broth (preferably homemade)

5 tablespoons (70 g) unsalted butter, divided

½ cup (80 g) finely chopped onion

1½ cups (300 g) Arborio or Carnaroli rice

½ cup (120 ml) dry white wine

½ cup (50 g) grated Parmigiano-Reggiano cheese

Pinch of kosher salt

A few gratings of black pepper

Cover the dried mushrooms in warm water and soak for 20 minutes, or until soft and pliable. Drain, but reserve the soaking liquid. Remove the stems, squeeze out any additional liquid, and slice the caps. (Save the stems for another use.) Strain and discard the debris from the soaking liquid.

Remove and save the stems from the fresh mushrooms, and thinly slice the caps.

Heat the broth and strained mushroom-soaking liquid in a 2-quart (2 L) pan to a slow simmer.

Melt 3 tablespoons (42 g) of the butter over medium heat in a heavy, 2½- to 3-quart (2.5 to 3 L) pan. Add the onions and cook for 3 to 4 minutes, until they are soft but not brown. Add all the mushrooms. Cook for an additional 3 to 5 minutes, stirring occasionally.

Increase the heat to high, add the rice, and stir to coat with the butter. Cook for 2 to 3 minutes. Add the wine and cook until it evaporates.

Add a ladle of hot broth and start stirring, preferably with a wooden spoon, making sure to include the rice on the sides and bottom of the pan. Keep stirring until all the liquid is absorbed. Then add another ladle, and repeat. Keep adding ladles of liquid for about 20 minutes, and then taste.

The rice should be al dente, just like well-cooked pasta: a little firm in the center. If it's more than a little firm, add more broth and cook and stir for an additional 5 minutes or so.

Remove the pan from the heat and stir in the cheese and the remaining 2 tablespoons (28 g) butter. Add the salt and pepper. Cover, let it rest for a few minutes, and then serve.

—

4 to 6 servings

NOTE

You can add a mix of agaricus, blewits, or wine caps in with the shiitake.

Braised Leek and Shiitake Gratin

There are hundreds of recipes combining mushrooms and this mild member of the onion family. Braised on top of the stove or roasted in the oven; leeks chopped, halved, or sliced; with cheese, with cream, with wine: almost any way you treat the two ingredients, they make an excellent match. Leeks are lovely, but can be really dirty! Be sure to clean them thoroughly before using them in any recipe. And use them you should; this gratin is an elegant side dish for any roast, be it lamb, chicken, beef, or whatever else you'll serve for a special dinner.

FOR BRAISE

4 leeks, trimmed and well cleaned

4 tablespoons (56 g) unsalted butter, divided

1 pound (454 g) shiitake mushrooms, stems removed and caps sliced

½ cup (120 ml) chicken broth, preferably low-salt

3 sprigs fresh thyme

¼ cup (60 ml) cream

Salt and pepper, to taste

1 teaspoon (5 ml) balsamic vinegar, or more if desired

FOR TOPPING

2 tablespoons (28 g) unsalted butter

½ cup (50 g) panko or other bread crumbs

TO MAKE THE BRAISE: Cut the leeks into 1-inch (2.5 cm) rounds. Set aside.

Heat 2 tablespoons (28 g) of the butter over medium heat in a large broiler-safe braising pan. When it stops foaming, add the mushrooms and sauté for about 2 minutes, until lightly browned, stirring occasionally. Be careful not to crowd the pan.

Remove the mushrooms and add the remaining 2 tablespoons (28 g) butter to the pan. Add the leeks and sauté for about 5 minutes, stirring occasionally.

Return the mushrooms to the pan, add the broth and thyme, and bring to a boil. Reduce the heat, cover, and simmer for 15 to 20 minutes, until the leeks are tender. Uncover, add the cream, and cook for a few minutes longer, until most of the liquid has been absorbed. Remove the thyme stems. Check and correct seasoning, adding salt and pepper if desired. Stir in the vinegar.

TO MAKE THE TOPPING: Preheat the broiler. Melt the butter in a small sauté pan and add the panko. Mix until the crumbs are coated. Spread the crumbs on top of the leeks and mushrooms and run the pan under the broiler briefly, until the top is browned and crisp.

—

4 to 6 servings

NOTE

You can substitute blewits for the shiitake. If you can do, use the whole mushroom, rather than removing the stem, and reduce the quantity to 10 ounces (280 g). You can use vegetable broth instead of chicken broth to make this dish vegetarian.

Beef Burgundy

There are probably as many variations of beef burgundy, a.k.a. boeuf bourguignon, as there are chefs. Julia Child herself had at least four! Each is a little different, but similar as well. They require an inexpensive cut of stew meat, onions, and of course, red wine. Most agree there should be mushrooms. We'll use agaricus, but wine caps would be great, too. This recipe is adapted from one of Julia Child's many versions, complete with scads of her signature ingredient: butter. That's how she liked to cook, and your guests will be the happy beneficiaries!

FOR STEW

1 tablespoon (15 ml) olive oil

6 ounces (170 g) bacon, cut into small sticks ¼ by 1½ inches (6 mm by 4 cm)

3½ to 4 pounds (1.6 to 1.8 kg) beef chuck, trimmed and cut into 2-inch (5 cm) chunks

2 carrots, thinly sliced

1 onion, thinly sliced

1 (750 ml) bottle dry red wine, divided

Salt and pepper, to taste

4 sprigs fresh thyme

2 to 3 cups (470 to 700 ml) beef stock

1 tablespoon (15 g) tomato paste

2 tablespoons (30 g) unsalted butter

2 tablespoons (16 g) flour

FOR ONIONS

2 tablespoons (30 g) unsalted butter

16 to 20 small white onions, peeled

1 to 2 cups (235 to 470 ml) water

FOR MUSHROOMS

2 tablespoons (30 g) unsalted butter

1 pound (454 g) agaricus or wine cap mushrooms, quartered

TO MAKE THE STEW: Preheat the oven to 325°F (170°C, or gas mark 3).

Heat the oil in a Dutch oven or other large casserole dish over medium heat. Add the bacon and sauté for 5 to 10 minutes, until lightly browned. Remove with a slotted spoon to a side dish, leaving the hot fat in the pan.

Turn the heat to medium-high. Dry the beef with paper towels. Brown in the hot fat, turning occasionally, for 5 minutes. Be careful not to crowd the pan; do this in several batches if necessary. Add the beef chunks to the side dish as they're done.

Add the carrots and onion to the pan and brown briefly, stirring.

Drain any remaining fat from the pan. Pour in 1 cup (235 ml) of the wine and deglaze the pan, cooking down the liquid and scraping up all the stuck-on bits.

Return the beef and bacon to the pan. Add the salt, pepper, and thyme. Mix the remaining wine and the stock with the tomato paste and pour over the beef until it is just about covered. Bring to a simmer, cover, and bake for 1½ to 2½ hours. Check after 1½ hours to see whether the meat is tender.

TO MAKE THE ONIONS: In a medium saucepan, melt the butter and sauté the onions briefly, and then add ½ inch (1.3 cm) of water. Cover and braise until tender, 20 to 30 minutes.

TO MAKE THE MUSHROOMS: In a medium skillet, heat the butter, add the mushrooms, and sauté until nicely browned, 5 to 10 minutes. Be careful not to crowd the pan.

When the meat is tender, add the mushrooms and onions to the stew.

Combine the 2 tablespoons (30 g) butter with the flour in a small bowl. Gradually whisk in a few tablespoons of the hot stew liquid, and then add back into the stew. Simmer for a few minutes to thicken. Make more flour-butter paste and add in the same manner if you'd like a thicker stew.

—

6 servings

Savory Mushroom Tart

This is quite a popular appetizer at Bistro 63 in Barronett, Wisconsin. Although many savory ingredients work well in a tart—think of squash, onions, chicken—executive chef Jeno Herman is an expert on edible wild mushrooms, and this recipe, adapted slightly from his, does a great job of showing off his expertise.

1 sheet frozen puff pastry, thawed but still cold

½ cup (35 g) shiitake mushrooms, stems removed

½ cup (35 g) oyster mushrooms

3 tablespoons (45 ml) extra-virgin olive oil

1 clove of garlic, minced

Salt and pepper, to taste

⅓ cup (80 ml) beef stock

3 tablespoons (45 ml) heavy cream

1 tablespoon (4 g) chopped fresh parsley

1 tablespoon (3 g) chopped fresh rosemary leaves

1 tablespoon (3 g) chopped fresh thyme leaves

3 tablespoons (27 g) crumbled goat cheese

Preheat the oven to 400°F (200°C, or gas mark 6). Line a baking sheet with parchment paper.

Unfold the pastry sheet and roll it out gently on a lightly floured surface. Cut into two long rectangles and place on the baking sheet, leaving space between them. Bake for 10 to 15 minutes, or until golden brown and puffy.

Chop the mushrooms. Heat the olive oil in a medium pan over high heat. Add the mushrooms and sauté until they are golden brown and the outsides are slightly hard.

Add the garlic, salt, and pepper and sauté for 1 to 2 minutes. Do not let the garlic burn. Add the stock, cream, and herbs. Turn down the heat and simmer until reduced by half.

Pour the mushroom mixture over the pastry. Sprinkle the goat cheese on top. Serve.

—

2 servings

NOTE

You don't have to stick to rectangles; you can have fun making the tarts into any shapes you like. Just make sure to leave enough space on each for a tasty heap of mushrooms!

Closing Words

My fascination with fungi started very early in life. As long as I can remember, I loved looking at mushrooms. As I got older, I started harvesting wild edibles to share with my friends. I would trade for a dozen eggs, or a jar of maple syrup. Things that others were afraid to harvest on their own could be traded for things I could use.

I developed a reputation for being able to identify mushrooms and for the ability to bring safe mushrooms to the table for the locals in my small town. As word spread, more and more people were asking me for mushrooms. I decided that the woods could no longer offer me the payload I needed. The demand continued to rise, and I could only put on so many miles before the hours of the day ran out. I was falling short on orders. I had to do something, so I decided to start cultivating my own mushrooms. It was an adventure, to say the least. The first year, I did about 250 shiitake logs and eight wine cap beds. I misread the instructions for wine caps and grew them the wrong way. The wrong way turned out to be the right way, and my production was astounding. The first wine caps came in early July, and it was an exciting moment. Seeing those deep red caps poking through the straw was like seeing a firstborn child enter the world. I had succeeded past all those weeks of doubt. Waiting for a mushroom to grow had seemed like a futile eternity. Every day I had questioned the success of the project.

The shiitake mushrooms had given me little hope as well, until they started to fruit that fall. Another sign of possible success had shown its face. This would be enough to supply everyone in the area who wanted mushrooms, I thought. It proved to be false on a massive scale.

I now have 8,500 producing logs with more on the way, a grow room capable of growing 200 pounds (91 kg) of oysters per week, and wine cap beds everywhere there is room for them. I have several area restaurants that use my mushrooms, and two farmers' markets; they've proved to me what an immense calling there is for fresh mushrooms in the area.

I am up to my ears in mushrooms every day of the year. I have never lost any interest in fungi, and every time I walk out into the mushroom farm, I am still fascinated, just as I was as a child. I can watch a mushroom grow from nothing. Its beauty is unique, and no other mushroom will look exactly like that one. Wild mushrooms are still my passion, but the cultivated mushrooms are to me what most people would call their pets. They are a domesticated version of wildlife that I share my space with.

—*Tavis Lynch*

Tavis's farm

List of Suppliers

Field and Forest Products
Peshtigo, WI
fieldforest.net
800-792-6220

Fungi Ally
Hadley, MA
fungially.com
978-844-1811

Fungi Perfecti
Olympia, WA
fungi.com
800-780-9126

Midwest Grow Kits
Algonquin, IL
midwestgrowkits.com
800-921-4717

Mushroom Mountain
Easley, CA
https://mushroommountain.com
864-855-2469

Smugtown Mushrooms
Rochester, NY
smugtownmushrooms.com
585-690-1926

Stockport Gourmet Mushrooms
Heath, OH
740-323-0793

Superior Mushrooms, LLC
Mason, WI
superiormushroomllc@gmail.com
715-413-1341

Sylvan Inc.
Sixteen locations serving more than sixty-five countries
sylvaninc.com

Where to Get Even More Information

Australasian Mycological Society
australasianmycologicalsociety.com

Beginning Farmers
beginningfarmers.org

European Mycological Association
www.euromould.org

Fungal Network of New Zealand
New Zealand Mycological Society
funnz.org.nz

FUNGI Magazine
www.fungimag.com

The Fungi of California
mykoweb.com/CAF

Mushroom Expert
mushroomexpert.com

The Mushroom Growers' Newsletter
mushroomcompany.com

North American Mycological Association
namyco.org

Northeast Mycological Federation
nemf.org

ShiiGA (Shiitake Growers Association)
shiigaw.org

United Kingdom
fungus.org.uk

About the Author

Tavis Lynch is an amateur mycologist from northwestern Wisconsin. He has spent his entire life in the woods, learning about the fungi of the North Woods. He teaches several mycology classes for Wisconsin Indianhead Technical College and for the University of Wisconsin. During the warmer months, he leads public and private forays for people to learn and to collect edible wild mushrooms. He also does daily farm tours and helps new cultivators get started growing and marketing mushrooms. His farm produces thousands of pounds of exotic mushrooms each year that are sold to local restaurants and directly to the public. He is active in the Shiitake Growers Association (ShiiGA), of which he is currently president.

Index

Printed in the USA
CPSIA information can be obtained
at www.ICGtesting.com
JSHW070524181124
73462JS00003B/3

9 780760 390788